RUSSIAN 'HYBRID WARFARE'

OFER FRIDMAN

Russian 'Hybrid Warfare'

Resurgence and Politicisation

HURST & COMPANY, LONDON

First published in the United Kingdom in 2018 by
C. Hurst & Co. (Publishers) Ltd.,
41 Great Russell Street, London, WC1B 3PL
© Ofer Fridman, 2018
All rights reserved.

The right of Ofer Fridman to be identified as the author of this publication is asserted by him in accordance with the Copyright, Designs and Patents Act, 1988.

A Cataloguing-in-Publication data record for this book is available from the British Library.

ISBN: 9781849048811

This book is printed using paper from registered sustainable and managed sources.

www.hurstpublishers.com

Printed and bound in Great Britain by Bell & Bain Ltd, Glasgow

But if the thought corrupts language, language can also corrupt thought. A bad usage can spread by tradition and imitation even among people who should and do know better.

George Orwell, 1946

CONTENTS

Acknowledgements ix
Timeline and Glossary xi

Introduction 1

PART 1
THE RISE OF 'HYBRID WARFARE'

1. The Conceptual Foundations of 'Hybrid Warfare' 11
2. The Birth of 'Hybrid Warfare' 31

PART 2
THE RISE OF *GIBRIDNAYA VOYNA*: THE RUSSIAN THEORY OF HYBRID WARFARE

3. Reading Evgeny Messner: The Theory of 'Subversion-War' (*Myatezhevoyna*) 49
4. Net-Centric and Information Wars: Modern Theories of Subversion 75
5. The Rise of *Gibridnaya Voyna* 91

PART 3
THE POLITICISATION OF HYBRID WAR AND *GIBRIDNAYA VOYNA*

6. 'The Russians Are Coming': The Politicisation of Russian Hybrid Warfare 101

CONTENTS

7. '*Gibridnaya Voyna* against Us Is Coming': The Politicisation of Western *Gibridnaya Voyna* — 127
Conclusions: The Rise of Russian 'Hybrid War'—Lessons For the West — 153
P.S. A Lesson for Russia — 177

Notes — 179
Index — 225

ACKNOWLEDGEMENTS

I am indebted to many scholars whose works are dutifully noted throughout the book; they might not subscribe to the argument I present and definitely bear no responsibility for any errors and mistakes in the text. The responsibility for all views and mistakes is mine and mine alone.

This project would not have been possible without the generous and priceless assistance of many people, to whom I would like to express my gratitude. On the academic front, special thanks are due to Beatrice Heuser, to whom I owe a great deal. I was fortunate to benefit from her approachability and encyclopaedic knowledge; her remarks and continuous support were absolutely invaluable. For his comments and advice during the initial stages of my research, I am especially grateful to Frank G. Hoffman, whose help and support encouraged me to carry on with my research. Thanks are also due to Derek M.C. Yuen for his useful comments on my interpretation of 'unrestricted warfare'.

In the process of completing this project, I was fortunate to spend a highly productive time at the King's Centre for Strategic Communications (KCSC) at King's College London. For making my time especially fruitful, I owe my appreciation to Neville Bolt, whose readiness to help went far beyond my expectations; to David Betz, whose wise advice I benefited from enormously; and to the remaining members of the centre's team, whose valuable comments were always useful. I also want to acknowledge Sagit Carmeli for her devoted and excellent editing, which turned my scrambled thoughts into a readable text—this project would not have been possible without you.

I would also like to express my gratitude to friends and relatives, of whom there are too many to list, who had probably never quite understood why, or indeed what, I was doing, but always believed in me and in my project nevertheless. I thank you all for believing in me even when I did not.

ACKNOWLEDGEMENTS

Last, but definitely not least, I would like to thank the Gerda Henkel Foundation for supporting this project, and Thomas Podranski, manager of the special programme entitled 'Security, Society and the State', who helped me through many intricate administrative issues.

TIMELINE AND GLOSSARY

Timeline and Glossary*

The West

Political Warfare
The use of political means to compel an opponent to do one's will based on hostile intent.

Fourth Generation Warfare
A type of warfare that is an evolved form of insurgency, and is rooted in the fundamental precept that superior political will, when properly employed, can defeat greater economic and military power. It makes use of societal networks to carry on its fight by directly attacking the minds of enemy's decision-makers.

Net-Centric Warfare (NCW)
NCW focuses on the combat power that can be generated from the effective linking or networking of the warfighting enterprise. It is characterised by the ability of geographically dispersed forces to create a high level of shared battlespace awareness that can be exploited via self-synchronisation and other network-centric operations to achieve commanders' intent.

Unrestricted warfare (Originally developed in China)
Warfare based on combinations that transcend technical, scientific, theoretical, psychological, ethical, traditional, customary, and other sorts of boundaries, thus erasing the boundary between the battlefield and what is not the battlefield and what is a weapon and what is not, between soldier and non-combatant, between state and non-state and super-state.

Compound Warfare
A simultaneous use of a regular or main force and an irregular or guerrilla force against an enemy, when the whole is greater than the sum of the parts.

Hybrid Warfare (Hoffman)
Incorporates a range of different modes of warfare including conventional capabilities, irregular tactics and formations, terrorist acts including indiscriminate violence and coercion, and criminal disorder.

Hybrid Threats (NATO)
Multidimensional adversaries that employ a complex blend of means that includes the orchestration of diplomacy, political interaction, humanitarian aid, social pressures, economic development, savvy use of the media, and military force.

'Russian Hybrid Warfare' (NATO)
The use of military and non-military tools in an integrated campaign designed to achieve surprise, seize the initiative and gain psychological as well as physical advantages utilising diplomatic means; sophisticated and rapid information, electronic and cyber operations; covert and occasionally overt military and intelligence action; and economic pressure.

Russia

Active Measures (not present in contemporary discourse)
Overt or covert activities conducted by intelligence agencies to influence political interests of adversary countries, undermining their foreign policy, subverting and weakening their positions, and disrupting their plans and goals.

Subversion-War (Evgeny Messner) (Unavailable in the Soviet Union)
Direct or indirect utilisation of already existing domestic political, social, economic and other turmoil, created by accelerated political-socio-cultural transformations, for political benefits.

Information War (Igor Panarin)
A confrontation between parties, represented by the use of special (political, economic, diplomatic, military and other) methods based on different ways and means that influence the informational environment of the opposing party in order to achieve clearly defined goals.

Revival of Subversion-War

Net-Centric War (Alexander Dugin)
A fight for the control over an informational dimension network in which major strategic operations are developed, including mass media, financial instruments, access to techologies, political and cultural elite, centres of active youths, etc.

'New-Generation War' (Chekinov and Bogdanov)
A war, in which a leading role is taken by the information-psychological struggle, directed to achieve superiority in the sphere of command and control, as well as to suppress the morale of the military personnel and the population of the adversary.

Theory of 'Controlled Chaos'
Reforming the mass consciousness, worldviews, and the spiritual sphere for political purposes by subjecting individuals to modern means of manipulation.

'Colour Revolution'
A change of the legitimate regime in an adversary country by destabilising the political situation inside the country with the predominant use of non-military methods of struggle.

Gibridnaya Voyna
An amplification of any possible social, political, economic or ideological divisions in the adversary's society that would help to undermine its political, economic and military cohesion and resilience.

Timeline
The Cold War | 1990s | 2000s | 2010s

---

* The appearance of concepts on the timeline represents the period when they were developed and not necessarily their period of influence.

xi

INTRODUCTION

I became interested in so-called hybrid warfare in late 2015, when two different books navigated their way on to my table. The first was entitled *NATO's Response to Hybrid Threats*, edited by two researchers from the NATO Defence College in Rome, Guillaume Lasconjarias and Jeffrey Larsen;[1] the second was entitled *'Hybrid Wars' in the Chaotic World of the 21st Century* (*'Gibridnyye Voyny' v khaotiziruyushchemsya mire XXI veka*), edited by Pavel Tsygankov, a professor at Lomonosov Moscow State University.[2] Since the first of the two books was written in the West and the second in Russia, I expected to find different views and interpretations of more or less similar ideas, particularly so as both books claimed to deal with the same phenomenon—hybrid war. However, when reading these books, I was surprised to discover that the only mutual ground between them was their titles.

Intrigued, I went on a quest—a research journey into the blue waters of academic, military and political scholarship on the nature of hybridity in contemporary conflicts in Russia and the West. This journey led me to study many different conceptualisations and perspectives, some of which were more interesting than others; some were intellectually challenging, while others tried to reinvent the wheel; some were academically solid and well researched, whereas others were filled with contradictions.

As I continued my research, it quickly became clear that the concept of 'hybrid warfare' had been politicised, both in the West and in Russia. A concept that had initially been intended to offer a better understanding of the nature of contemporary conflicts has been weaponised, becoming a tool in internal manoeuvring for finance, public opinion and political power in Russia and the West, as well as a means of intimidation in relations between the two.

RUSSIAN 'HYBRID WARFARE'

In their analysis of Russia's reaction to the Ukrainian Crisis in 2014, Western experts have argued that the Kremlin has embraced hybrid warfare as its strategy in Ukraine. Many Russian strategists and political observers, in contrast, have contended that the West is in fact conducting a hybrid war (*gibridnaya voyna*) against Russia. This book aims to explain the reasons behind these mutual criminations. What does hybrid warfare mean in the West, and what is *gibridnaya voyna* in Russian eyes? Are they similar, and if not, how do they differ? Why have the terms hybrid warfare and *gibridnaya voyna* become so politicised in Russia and the West since the Ukrainian Crisis? Did the Kremlin, or the West, behave in a way that requires a new conceptualisation of warfare, or is the term itself just a new label for an old concept?

Answering these questions was the principal focus of my research. Yet the book does not focus on the Ukrainian Crisis per se. Instead, it examines the fundamental differences between hybrid warfare, *gibridnaya voyna* and all other associated concepts that preceded, accompanied and outlived these two in the Western and Russian scholarship. This book tries to grasp how political forces have shaped conceptual thinking, which has made it easier for one side to accuse the other of illegal actions while, simultaneously, enabling one side to fend off their adversary's criticism of any wrongdoings.

What this book is about (and what it is not about)

This book is not about hybrid war as an actual practice. The book discusses neither the tactics of what Western analysts commonly call Russian hybrid, cyber or information wars (with their hacks, leaks, internet trolls or disinformation and propaganda operations conducted by pro-Russian or Russian-financed media), nor the actions of what Russian experts describe as the Western *gibridnaya voyna*. It is not to say that these are unimportant; they obviously are. But examining them is similar to analysing different weapons and the methods of their employment, rather than investigating the actual nature of the conflict, how it is conceptualised by each side and where these understandings come from—two equally important, yet quite different tasks.

Since this book deals with the discourse on and perceptions of the ongoing enmity between Russia and the West, which is conceptualised and contextualised by the language of hybridity, this book is, after all, about hybrid war. There is a very thin line between how reality is perceived and reality itself, and since language is the only way to describe them both, the discourse surrounding the concept serves as a way to cross this line. Without falling into

INTRODUCTION

Foucauldian patterns of interpretation, it is difficult to escape the fact that certain political actors, on both sides, have benefited from the ongoing discourse of hostility. Defence budgets have risen in Russia and in NATO; Sweden, which is not a member of NATO, has reintroduced conscription; NATO is currently deploying forces in Eastern Europe on the largest scale since the end of the Cold War; and the Kremlin has renewed strategic bomber flights, while also introducing new pieces of military hardware on an almost monthly basis.

Neither Russia nor the West claims to want a repetition of the Cold War. Yet both sides seem to be talking themselves into exactly such a scenario—when analysing the language both sides use to describe the reality they find themselves in (or their perceptions of it), it is difficult to avoid the conclusion that a new title for a similar form of confrontation (hybrid war or *gibridnaya voyna*) does not really change its nature. Today's confrontations are undoubtedly very different from what the world experienced during the Cold War; but the Second World War was very different from the First World War: the weapons were more destructive, the tactics were more sophisticated and the outcomes more devastating. And yet the nature of both wars was ultimately very similar: they were both total wars fought, to some extent, by the same parties.

In analysing the ways in which Russian hybrid warfare has been conceptualised in the West, and how *gibridnaya voyna* has been perceived in Russia, this book seeks to explore the nature of the confrontation between the two, rather than the weapons and tactics involved, as only a true understanding of its nature will allow for any certainty with regard to whether hybrid war is simply a second Cold War, or, indeed, something entirely new.

Unfortunately, it seems that the former is closer to the truth—the West is again preoccupied by a 'red under every bed' mentality and the Kremlin is chasing after new *vragi naroda* (enemies of the people). Ultimately, however, whether the confrontation between the West and Russia is called a second Cold War or a hybrid war is largely irrelevant; what matters is the language used to describe this confrontation, because it is language that turns our perceptions into reality, and it is this that is the subject of the current book.

Finally, it should also be noted that this book does not focus on the Soviet concept of 'active measures' (*aktivinye meropriatia*)—'the heart and soul of the Soviet intelligence'—which, according to retired KGB Major General Oleg Kalugin, were intended:

> to weaken the West, to drive wedges in the Western community alliances of all sorts, particularly NATO, to sow discord among allies, to weaken the United

3

States in the eyes of the people of Europe, Asia, Africa, Latin America, and thus to prepare ground in case the war really occurs. To make America more vulnerable to the anger and distrust of other peoples.[3]

There are two main reasons for this. The first is the fact that active measures have already been widely discussed in the West.[4] The second and more important is that the concept of active measures has largely been ignored in contemporary Russian academic, military and political discourse. There are several possible reasons for this: from an attempt to avoid an unhelpful association with the Soviet past, through a sincere disappointment with concepts that failed during the Cold War, to a genuine attempt to build something new and more powerful.

While many readers will find conceptual similarities between the Soviet concept of active measures and the more contemporary Russian theories and ideas that are discussed in this book, creating any conceptual linkage is at the reader's peril. In doing so, the reader should remember that while some Russian actions can be conceptually described as an adaptation of active measures to twenty-first-century realities, the differences between them are similar to the differences between the means and methods of the First and the Second World Wars. And any attempt to comprehend Russia's use of non-military means and methods in the twenty-first century through the prism of active measures would be equivalent to an attempt of a British officer from the Somme in 1916 to grasp the realities of the Second Battle of Sedan in 1940.

On the one hand, there is little doubt that there are certain similarities between active measures as a means to achieve political goals and the different methods at the heart of the Russian theory of *gibridnaya voyna*. On the other hand, active measures, as a concept, has not shaped Russian thinking about twenty-first-century conflicts, and since this book focuses more on contemporary discourse than on historical background, a discussion of active measures is not relevant to the matter at hand.

A note on translation

Half of this book focuses on the Russian side of the discourse surrounding hybrid warfare. As such, it is based on numerous sources that have never been translated to English before. Russian, like any other language, is deeply embedded in its own unique historical–cultural context, and like any translator, I had to navigate the painful dilemma of being caught between following the words verbatim or trying to translate their proper meaning. One of the best examples of this dilemma is discussed in Chapter 4, when the narrative turns to the writings of military theorist Evgeny Messner and his theory of subversion-war. Moreover, sometimes the intricate particularities of each language simply prevent the translator from bringing the reader the true meaning of a given text. *Russkiy* and *rossiyskiy*, for instance, are both translated to English as 'Russian', but there is a vast difference between *russkaya zemlya* (Russian land) and *rosiiyskaya federatziya* (Russian Federation) because *russkiy* describes something that belongs to the Russian ethnicity, whereas *rossiyskiy* describes something belonging to the Russian state.

Following the meaning of the text, rather than its direct translation, can prove dangerous, as the translator will inevitably run the risk of putting their own words into the original author's mouth. But a direct word-by-word translation is no less dangerous, as it risks losing the meaning altogether. Ultimately, and despite the risks involved, I elected to choose the first of these two approaches, and I take full responsibility for any mistakes and misinterpretations. My only hope is that my awareness of the respective pitfalls of both approaches has kept my mistakes to a minimum.

On the structure of this book

The relationship between reality and the concepts used to explain and describe it results from a series of highly complex interactions—real political events produce new conceptual visions, which in turn shape new political discourses, which then create new political actions, which then generate newer concepts, and so on and so forth. Thus any attempt to examine this relationship requires a narrow prism of analysis that is able to capture this nexus, with the aim of producing practical recommendations. As a result, the full story of Russia's relationship with the West is much more complicated than the snapshot this book provides. The relationship itself consists of four different aspects: (1) real events during the Cold War and the different concepts used to describe them; (2) political–military events since the end of the Cold War; (3) the concepts used to describe and explain these events in the post-Cold War geopolitical environment; and (4) how these concepts have shaped ongoing political events. Covering this entire chain of events is quite simply beyond the scope of this book—the narrative therefore focuses merely on the final step in the relationship between events and conceptualisation: in other words, how theoretical concepts shape the political–military behaviour and actions of Russia and the West.

Moreover, it is important to remember that the discussion of how post-Cold War political–military events have shaped the theoretical conceptualisation of warfare has already been thoroughly discussed in the literature dealing with the contemporary conceptualisation of warfare (this literature is discussed in the first and second parts of the book).

Accordingly, this book deliberately avoids covering how real world events have shaped political–military concepts. Instead, it explicitly focuses on how the conceptualisation of warfare in Russia and the West has shaped contemporary political events; this is particularly important given the lacuna in the existing literature on the relationship between conceptual military thinking and contemporary Russian–Western relations.

This book attempts to grasp two relatively independent narratives of thinking about conflicts that have a common beginning, namely the end of the Cold War, completely different ways of thinking about conflicts (Part 1 for the Western way and Part 2 for the Russian way), and an ending, where they finally meet and engage with each other (Part 3). Hence the book is less about how Western and Russian strategists have sought to conceptualise war in the post-Cold War environment and more about how the Western and Russian conceptualisations of war have been politicised to shape the contemporary relations between the two. It goes without saying that the discussion of the former (presented in Parts 1 and 2) is essential for understanding the latter (Part 3), however it serves as an important by-product, not the true goal of this book.

Carl von Clausewitz once argued that 'War is nothing but a duel', underlining the fact that any conflict involves at least two parties.[5] Since the parties involved in a given conflict constantly interact with each other, it is difficult to separate the actions of one from the other. There are two ways to address this problem: either to follow the story, constantly referring to the actions and counteractions taken by each party, or to focus separately on each side and trust the reader to assemble the puzzle. Since the story this book tries to outline is not about battlefields (whether real or virtual), but rather about ideas, the book follows the second path and focuses first on the West's side of the story before discussing the Russian one. In doing so, I was aware that some readers might find it irritating, especially in Part III that focuses on the politicisation processes, which, by their very nature, are results of consecutive interactions between the involved parties and, therefore, are difficult for comprehension without the context of the adversary's actions. I trust my readers, however, to keep in mind not only that each chapter presents only one

INTRODUCTION

side of the story but also that the following chapters would help in the understanding of the previous ones, as only all the pieces together can uncover the complete picture of the puzzle.

Part 1: The rise of 'hybrid warfare'

Part 1 of the book examines hybrid warfare as it was originally conceptualised by Frank Hoffman. Since hybrid warfare is a product of US military thought, it is important to understand not only the US concept but also the process by which it was developed. The book's first chapter consequently focuses on the theoretical debate within the US military that led Hoffman to develop his theory. Following an analysis of Hoffman's works, as well as those of other US military thinkers, the second chapter examines these works in the context of US strategic culture. This serves to provide a more complete overview of the original concept and its characteristics in US military discourse. As the term 'hybrid warfare' was invented in the United States, and was only later adopted by the Russians or reconceptualised by European scholars and military thinkers, these chapters are vital for understanding the initial meaning of hybrid warfare.

Part 2: The rise of gibridnaya voyna: *the Russian theory of hybrid warfare*

When analysing the Russian literature on the phenomenon of *gibridnaya voyna*, it quickly becomes clear that the concept is completely different from the US concept of hybrid warfare, which focuses chiefly on military tactics and operational activities. According to Russian specialists, the main purpose of *gibridnaya voyna* is to avoid the traditional battlefield with the aim of destroying the adversary via a mixture of ideological, informational, financial, political and economic methods, ultimately leading to socio-cultural disintegration and, eventually, social collapse. This idea of using non-military means to undermine an adversary is certainly not new in Russian academic, political and military discourse, as its conceptual origins can be traced to the mid-twentieth-century writings of Evgeny Messner on the theory of 'subversion-war', as well as the work of two Russian scholars, Aleksandr Dugin and Igor Panarin, who conceptualised and promoted two different theories of non-military conflicts in the 1990s and early 2000s: 'net-centric war' and 'information war'. This part of the book is made up of three chapters. While the first two chapters provide a comprehensive analysis of the different theories that preceded the concept of *gibridnaya voyna* and served as the conceptual basis for its creation, the third chapter focuses on the concept itself and its main characteristics.

Part 3: The politicisation of hybrid warfare and gibridnaya voyna

The third part of the book consists of two chapters. Whereas the first chapter focuses on the conceptual and political debate surrounding Russian hybrid warfare in the West, the second chapter analyses the Russian discourse on the role and place of non-military means and methods in contemporary conflicts and the *gibridnaya voyna* that the West, according to the Russians, have been waging against the Kremlin. As noted above, this division of the story into two separate chapters is largely artificial, as each one of the discourses was developed in context of the other. Yet this was the only possible way to analyse each of the cases and hence provide a better understanding of who stands behind the politicisation of hybrid warfare/*gibridnaya voyna*, what they are trying to achieve, and how successful their actions have been.

PART 1

THE RISE OF 'HYBRID WARFARE'

1

THE CONCEPTUAL FOUNDATIONS OF 'HYBRID WARFARE'

In the literature produced in the West, the concept of hybrid warfare is most often associated with US military theorist Frank Hoffman. Beginning in the mid-2000s, Hoffman attempted to bridge the gap between the linear characterisation of (regular or irregular) warfare in the context of the twenty-first-century operational environment. Generalising from the experience of the Israel Defense Forces (IDF) with Hezbollah in Lebanon in 2006, Hoffman argued that: 'The blurring of modes of war, the blurring of who fights, and what technologies are brought to bear, produces a wide range of variety and complexity that we call Hybrid Warfare.'[1] According to Hoffman, hybrid warfare can be conducted by states and non-state actors and involves: 'A range of different modes of warfare, including conventional capabilities, irregular tactics and formations, terrorist acts including indiscriminate violence and coercion, and criminal disorder.'[2]

Hoffman was not the first to notice these developments—as he has stated, previous theories and observations heavily influenced his own theory. Indeed, a closer examination of the literature published before the emergence of Hoffman's concept of hybrid warfare suggests that the term 'hybrid' had already been used to describe the blurred line between regular and irregular forces and capabilities.[3] Yet Hoffman's work is nonetheless significant because it initiated the intellectual debate in the West over contemporary hybrid threats in which the concept of hybrid warfare has been seen as a new way to conceptualise twenty-first-century conflicts.

RUSSIAN 'HYBRID WARFARE'

This part of the book examines hybrid warfare as it was originally conceptualised and understood in the West. Since the concept is the product of US military thought, it is important to examine the context in which it was developed, in addition to the way it has been used by US scholars and members of the military. This chapter consequently introduces four main sources that influenced the idea of hybrid warfare: namely the concepts of unrestricted warfare, fourth generation warfare (4GW), compound warfare and the ideas outlined in the 2005 US National Defence Strategy.

The theory of unrestricted warfare

In February 1999, two Chinese People's Liberation Army (PLA) Air Force colonels, Qiao Liang and Wang Xiangsui, published a book entitled *Unrestricted Warfare*.[4] In China, the book became a bestseller, one that has been widely read at the top echelons of the PLA and Chinese Communist Party.[5] However, in the United States, the book was met with alarm. As Admiral Thomas Moorer, former chairman of the Joint Chiefs of Staff, put it: 'You need to read *Unrestricted Warfare* because it reveals China's game plan in its coming war with America ... China thinks it can destroy America by using these tactics.'[6]

As originally interpreted by US scholars and military strategists, Qiao and Wang's book promoted the immoral and potentially illegal transformation of warfare, in which the international system and international law 'will be ... required to be overthrown in order for the Chinese to achieve their desired policies'.[7] However, this interpretation rests on a misunderstanding of the book's title rather than the general concept, as the concept of unrestricted warfare has since ignited constructive debate in US military circles.[8]

Indeed, the original title of the book (超限战) may have been translated too literally; a less incendiary alternative would be 'warfare beyond bounds', and a more appropriate translation for the concept itself, based on the main narrative of the book, would be 'warfare that transcends boundaries'. In Qiao and Wang's words: 'Thus, we obtain a complete concept, a completely new method of warfare called "modified combined war that goes beyond limits".'[9]

As 'transcending boundaries' does not involve the negative connotations associated with the idea of 'unrestricted' warfare, such a translation would perhaps have avoided the furore that the book generated when it was first published. This becomes especially clear from a passage in the following section of the book: 'Unlimited surpassing of limits is impossible to achieve. Any

THE CONCEPTUAL FOUNDATIONS OF 'HYBRID WARFARE'

surpassing of limits can only be done within certain restrictions. That is, "going beyond limits" certainly does not equate to "no limits", only to the expansion of "limited".[10]

Thus, leaving aside the negative reception the concept of unrestricted warfare initially received, it has since gained a wider appreciation in the United States not least because, as Hoffman puts it, 'a closer reading of the text reveals a lot of useful and even obvious conclusions'.[11]

The book's authors make three contributions to the field of strategy: (1) a constructive observation of the phenomenon of war; (2) a comprehensive analysis of the transformation of warfare at the end of the twentieth century and the Gulf War against Iraq in 1991 as a symbol of this transformation; and (3) an intellectual conceptualisation of future conflicts in the twenty-first century.

While most of Qiao and Wang's work discusses the nature of war in the late twentieth century and makes some conceptual predictions about the future, the authors also make several important observations on the nature of war and its principles, as 'regardless of the form the violence takes, war is war, and a change in the external appearance does not keep any war from abiding by the principles of war'.[12]

The most prominent principle is the importance of combination— 'combining two or more battlefield factors together'[13]—for achieving victory over an adversary. Observing different examples from Chinese and Western military history, Qiao and Wang conclude that: 'Regardless of whether the war was 3,000 years ago or at the end of the 20th century, it seems that all of the victories display one common phenomenon: the winner is the one who combined well.'[14]

In other words, one of the most important factors to achieve victory in a war is the ability of strategists to combine different technologies, concepts of operations, means and methods in a way that offers significant advantages and improves existing warfare to a significant degree.[15]

In developing this idea of combination, Qiao and Wang correctly point out that not every combination is a potential force multiplier, but also that 'without understanding the secret of how to conduct combination, it will be useless to conduct combination 100 times incompetently'. Highlighting different historical examples, the authors suggest two main rules that contribute to the optimum combination: the rule of the golden section (0.618:1 ratio) and the side-principle rule (taken from linguistics, where one word modifies another and determines its tendency and features). The first rule was taken from the world of art and the second from Chinese grammar. Qiao and Wang successfully demonstrate the relevance of these rules to the phenomenon of war by asking 'if too many accidents demonstrate the same phenomenon, can you still calmly view them as accidents?' In the discussion that defines and fuses these two rules, Qiao and Wang advocate a principle according to which the combination should be applied in warfare. First, they argue that it is important to define a dominant element in each of the five main components of war: weapons, means, force, direction and sphere. Secondly, the relationships between the dominant weapons and all the weapons, the dominant means and all the means, the dominant force and all the forces, should be defined according to the 0.618:1 ratio.[16]

Qiao and Wang do not claim to have found a formula that would guarantee victory in all forms of war; instead, they claim to have identified a principle that increases the likelihood of winning a war: 'the key [to victory] is to grasp the essence and apply the principle', rather than act strictly according to its rule, because 'correct rules do not guarantee that there will always be victories; the secret to victory is to correctly apply rules'.[17]

As well as claiming that the principle of combination has played a vital role in achieving victory since the dawn of human conflict, Qiao and Wang also attempt to establish the typical combination behind these victories: 'Whoever [will be] able to mix a tasty and unique cocktail for the future banquet of war will ultimately be able to wear the laurels of success on his own head.'[18]

Analysing the impact of technological progress and globalisation processes on the nature of war, Qiao and Wang come to the conclusion that 'a war which changed the world ultimately changed war itself'.[19] These two developments not only lie at the core of the transformation of warfare at the end of the twentieth century but they are also strongly interwoven, mutually increasing the impact of each other, thus increasing the pace at which war has been transformed.

Building upon their understanding of success in war as an ability to produce successful combinations, Qiao and Wang claim that technological integration and globalisation, which marked the second half of the twentieth century, have created an unprecedented demand for more complicated combinations than ever before. According to their theory of unrestricted warfare, the twentieth-century's technological and geopolitical realities have created a situation that requires strategists to cultivate and implement what they call 'combinations that transcend boundaries'.[20] In terms of technological progress and globalisation, Qiao and Wang identify two separate but interconnected developments. On the one hand, they highlight the increasing role of information technology by claiming that:

THE CONCEPTUAL FOUNDATIONS OF 'HYBRID WARFARE'

[It is pointless for military organisations] to wrack their brains over whether or not information technology will grow strong and unruly today, because it itself is a synthesis of other technologies, and its first appearance and every step forward are all a process of blending with other technologies, so that it is part of them, and they are part of it, and this is precisely the most fundamental characteristic of the age of technological integration and globalisation.[21]

On the other hand, when analysing the influence of globalisation in this age of information technology, Qiao and Wang state that: 'Non-professional warriors and non-state organisations are posing a greater and greater threat to sovereign nations, making these warriors and organisations more and more serious adversaries for every professional army.'[22]

These two developments, according to the theory of unrestricted warfare, increase the role of what the authors call 'non-military' means and methods of warfare, including acts of terrorism, cyber-crimes and financial manipulations.[23]

It was these two observations that led Qiao and Wang to develop a strategic concept of 'combinations that transcend boundaries', based on the claim that modern warfare blurs 'technical, scientific, theoretical, psychological, ethical, traditional, customary, and other sorts of boundaries', thus erasing 'the boundary between the battlefield and what is not the battlefield, between what is a weapon and what is not, between soldier and non-combatant, between state and non-state and supra-state'. The authors consequently propose four new types of combinations, which, according to them, best characterise modern warfare. The first is 'supra-national combinations', which combine national, international and non-state organisations and 'assemble and blend together more means to resolve the problem in a range wider than the problem itself'. As modern countries are increasingly affected by international organisations (multinational, non-state, commercial, religious, criminal, terror, etc.), Qiao and Wang suggest that modern conflict has transcended the boundary of the nation state, and as 'the threats to modern nations come more and more often from supra-national powers, and not from one or two specific countries' there is 'no better means for countering such threats than the use of supra-national combinations'.[24]

The second type of combination is 'supra-domain combinations', which transcend the domains of battlefield. As modern information technology and globalisation impel 'politics, economics, the military, culture, diplomacy, and religion to overlap each other',[25] new forms of warfare have opened up in the domain of warfare, such as information warfare, financial warfare, trade warfare, and psychological warfare. The combinations of these domains of warfare constitute 'supra-domain combinations' that direct efforts into certain dimensions that are the most favourable for the accomplishment of the objectives of a given conflict.[26] The third type of combination is 'supra-means combinations'—a mix of different means within each domain of warfare that creates the most desirable effect on an adversary.[27]

The last type of combination, and probably the most important, is 'supra-tier combinations', which combine all levels of conflict into a single campaign. Analysing the nature of contemporary conflicts, characterised by 'supra-national powers' that use 'supra-means' in 'supra-domains', Qiao and Wang state that the boundary between tactics, operations, strategy and grand-strategy has become blurred. In modern warfare, actions on a tactical level might have strategic consequences, and strategic aims might be achieved by tactical actions: 'Bin Laden used a tactical level method of only two truckloads of explosives and threatened U.S. national interests on strategic level, whereas the Americans can only achieve the strategic objective of protecting their own safety by carrying out tactical level retaliation against him.'[28]

Qiao and Wang use many historical examples to support their conceptualisation, yet the case of the 1991 Gulf War is their main point of reference:[29] 'When we attempt to use wars that have already occurred to discuss what constitutes war in the age of technical integration-globalisation, only "Desert Storm" can provide ready-made examples.'[30]

According to Qiao and Wang, the Gulf War 'concluded the old era and inaugurated the new one'[31] because it was the first war that truly transcended the boundaries of warfare. There were 'supra-national combinations', as the United States was highly successful in creating a coalition of over thirty countries, including some that were hostile to each other, as well as receiving the support of almost all the countries in the United Nations together with international and non-state organisations (e.g. the World Bank, the World Trade Organisation). There were 'supra-domain combinations' as 'the 42-day military action of Desert Storm was followed by eight continuous years of military pressure + economic blockade + weapons inspections'.[32] There were also 'supra-means combinations' as different means were employed in each of the utilised domains (e.g. the authors discuss the military and psychological domains in detail).[33] While the authors themselves do not provide any examples of 'supra-tier combinations' from the Gulf War, such combinations can easily be identified. The best example is the use of relatively precise ammunition: though this was tactical in nature, it also had a strategic impact via infor-

THE CONCEPTUAL FOUNDATIONS OF 'HYBRID WARFARE'

mation technology and the live broadcasting of exploding targets, thus influencing the eyes and ears of the entire world.

Observing geopolitical developments in the second part of the twentieth century, the authors suggest that 'while we are seeing a relative reduction in military violence, at the same time we definitely are seeing an increase in political, economic, and technological violence'.[34] In other words, unrestricted warfare is an attempt to advocate for the triumph of non-traditional domains of violence (i.e. 'non-military' or 'non-war' actions) over purely military means to achieve desirable objectives, on the battlefields of the late twentieth and early twenty-first centuries.

At the end of their book, Qiao and Wang propose eight principles that already characterise contemporary warfare and will have even more influence on the nature of future wars. The first is 'omnidirectionality'. According to this principle, future wars will be characterised by a lack of distinction between what is and is not the battlefield and will include not only traditional military domains (i.e. land, sea, air, etc.), but also social spaces, such as politics, economics, culture and the psyche of the nations involved (as well as neutral nations). The participants in these wars will have:

> To give all-round consideration to all factors related to 'this particular' war, and when observing the battlefield or a potential battlefield, designing plans, employing measures, and combining the use of all war resources which can be mobilised, to have a field of vision with no blind spots, a concept unhindered by obstacles, and an orientation with no blind angles.[35]

The second principle is 'synchrony', which implies that actions in future wars will need to be conducted simultaneously in different domains. According to Qiao and Wang, technological integration and globalisation will not only allow for combinations of different activities with different means in different domains but also more synchronised and simultaneous execution of these combinations. If, in the past, objectives had to be accomplished in stages 'through an accumulation of battles and campaigns', in the future they might be accomplished 'under conditions of simultaneous occurrence, simultaneous action, and simultaneous completion'.[36]

The third and fourth principles that would characterise future conflicts, according to Qiao and Wang, are 'limited objectives' and 'unlimited measures'. The former highlights the importance of setting explicit and practical objectives for future wars, as 'setting objectives, which exceed allowable limits of the measures available, will only lead to disastrous consequences'. The principle of unlimited measures is based on the idea that 'to accomplish some designated objectives, one can break through restrictions and select among various measures'.[37] While each of these two principles introduces nothing new into the nature of warfare, their combination shapes the main idea of unrestricted warfare—limited objectives achieved by unlimited measures.

The fifth concept is 'asymmetry'—exploiting 'an enemy's soft spots'. Although this principle again offers little that is novel to the nature of war, according to Qiao and Wang the shift of warfare from the traditional military domain towards non-military domains (i.e. politics, economy, culture) will increase the role of this principle, offering more opportunities to weaker actors and exposing the vulnerabilities of the stronger ones.[38]

The sixth principle is 'minimal consumption' according to 'the rational designation of objectives and the rational use of resources'. In future conflicts, Qiao and Wang maintain that the growing number of possible combinations of objectives and the different measures to accomplish them will ultimately increase the possibility of 'high consumption with low effectiveness', and, therefore, it will be required (more than ever before): 'To combine the superiorities of several kinds of combat resources in several kinds of areas to form up a completely new form of combat, accomplishing the objective while, at the same time, minimising consumption.'[39]

The seventh principle is 'multidimensional coordination'—the coordination and cooperation of all required means and measures in all required domains in order to accomplish the objective of war. Since unrestricted warfare assumes that not only military but any domain can become a battlefield, the participants in future wars should be 'inclined to understand multidimensional coordination, as the coordination of the military dimension with various other dimensions in the pursuit of a specific objective.'[40]

The final principle of future wars, as suggested by Qiao and Wang, is 'adjustment and control of the entire process'. While, again, this principle does not represent a new characteristic of warfare, implementing it will become more complicated in a war with an increasing number of possible combinations, unexpected links and influential factors: 'The ability of these factors to cloud the issues of war, and their intense influence on war, means that loss of control over any one link can be like the proverbial loss of a horseshoe nail which led to the loss of an entire war.'[41]

In conclusion, it is important to focus on three main aspects of this theory. The first and most important is the idea that human conflict transcends the limits of the traditional military domain, thus infiltrating other domains of human interaction (i.e. politics, economics, culture) on the sub-national,

national and supra-national levels. Consequently, Qiao and Wang argue that 'He, who wants to win today's war, or those of tomorrow ... must "combine" all of the resources of war [military and non-military] which he has it his disposal and use them as means to prosecute war.'[42]

The second aspect is the fact that the authors recognise that there is little new in their idea of 'combination', as they argue that 'Alexander the Great and the martial kings of the Zhou Dynasty never heard of cocktails [i.e. combinations], but they knew the value of the combined use of things.'[43] The main argument, however, is that in modern and future conflicts this old technique of combination will be used to combine actions, means and methods that have not previously been considered parts of warfare.

This leads directly to the third main argument made by unrestricted warfare—the two primary drivers behind the appearance of this 'Warfare that Transcends Boundaries' are the technological and social changes at the end of the twentieth century (i.e. globalisation). Responding to these changes, the new principle of war is not 'using armed force to compel the enemy to submit to one's will', but 'using all means, including armed force or non-military force, military and non-military, and lethal and non-lethal means to compel the enemy to accept one's interests.'[44]

The theory of 'fourth-generation warfare'

The concept of fourth-generation war (4GW) was first introduced by a group of US Marines officers, led by military strategist William Lind, in a 1989 article entitled 'The Changing Face of War: Into the Fourth Generation.'[45] The idea of 4GW has since gained in popularity in Western military circles, including the works of Martin van Creveld[46] and Samuel Huntington[47] and has attracted many direct proponents,[48] together with those opposed to the idea.[49]

According to the article's authors, analysis of modern warfare should begin with the 1648 Peace of Westphalia, the settlement that ended the Thirty Years' War and established the state's monopoly on war.[50] Since then, the nature of warfare has transformed via three main generations: (1) manpower, (2) firepower, (3) manoeuvre. In the late twentieth century, warfare developed into the fourth generation: 'an evolved form of insurgency that employs all available networks—political, economic, social, military—to convince an opponent's decision-makers that their strategic goals are either unachievable or too costly.'[51] Moreover, according to 4GW, these shifts in the nature of warfare were initiated not by explicit military factors but by a combination of 'major political, economic, social, and technical changes' that preceded each transformation.[52] Interestingly, without explicitly mentioning this, the proponents of 4GW echo the concept of 'Military Revolution' suggested by Michael Roberts in the late 1950s and developed further by other military historians and strategists, such as Clifford Rogers, Geoffrey Parker, Max Boot, Williamson Murray and others.[53]

According to Thomas Hammes, one of the most prominent advocates of 4GW, the peak of first-generation warfare was the Napoleonic Wars of the early nineteenth century. These wars were a culmination event that intersected several developments required for the transition from the medieval military of feudal knights to the Napoleonic *Grande Armée*. While the development of reliable firearms and effective artillery was the first and most obvious aspect that allowed for the culmination of the use of massed manpower, the most important developments were 'the political system, the wealth-generating national economies, the social structures, and technologies capable of sustaining the mass armies of the Napoleonic era'. In other words, the transformation of medieval warfare into the first generation of modern war took almost 200 years and was driven by 'significant change in the political, economic, social, and technological structures of the time'.[54]

There is little new in this observation, as similar ideas had been put forth by military historians several decades before the appearance of the 4GW concept.[55] However, the characteristics of the first generation of modern warfare given by Lind, Hammes and others are important for understanding the 4GW theory. The most prominent characteristic of first-generation warfare is 'line-and-column' tactics, whereby 'battles were formal and the battlefield was orderly'. According to Lind, the main relevance of this generation of warfare 'springs from the fact that the battlefield of order created a military culture of order', as 'most things that distinguish military from civilians—uniforms, saluting, careful gradation of rank—were products of the First Generation and intended to reinforce the culture of order'.[56]

The 'battlefield of order' culminated in the nineteenth century and then started to break down. In the period that followed, armies were increasingly motivated by nationalistic ideas, and the introduction of new military technologies, such as rifled muskets, breech-loading firearms and machine guns, made 'line-and-column' tactics obsolete. As such, 'the culture of order that was consistent with the environment in which it operated has become more and more at odds with it'.[57]

According to the advocates of 4GW theory, the onset of the second generation of warfare, much like the first generation of warfare, resulted from a combination of social, political and technological changes. On the one hand, the increased effectiveness of states' power, industrialisation, more technologically advanced weaponry, together with improvements in transportation, communication, agriculture and health, as well as a vast increase in population, allowed for the recruitment of enormous armies. On the other hand, the rise of nationalism, particularly in the period during and after the Napoleonic Wars, generated a patriotic impulse among the citizens of nation states, ultimately bringing millions of men to the colours; this period of second-generation warfare would eventually culminate in the catastrophic losses incurred during the First World War.[58]

The transition process from the first to the second generation of warfare did not take place rapidly; instead, it was a gradual transformation that occurred over several decades: 'From the early battles of the U.S. Civil War to the battlefields of South Africa and the trenches of Far East, war provided repeated, clear examples of the effect that political, economic, social, and technical changes having on warfare.'[59]

Whereas the first generation of warfare was shaped by the 'line-and-column' tactic that had required a mass deployment of manpower, second-generation warfare 'sought a solution in mass firepower, most of which was indirect artillery fire'. In other words, new military technologies led to two interconnected transformations: lateral dispersion of the forces on the battlefield and the ability to concentrate firepower on an unprecedented scale.[60] While the first change undermined the military culture of order that had been established during the previous generation of warfare, the second transformation not only preserved this military culture but attached an increasing level of importance to it. In Lind's words:

> Centrally controlled firepower was carefully synchronised (using detailed, specific plans and order) for infantry, tanks, and artillery in a 'conducted battle' where the commander was, in effect, the conductor of an orchestra. [Therefore] the focus was inward, on rules, processes, and procedure. Obedience was more important than initiative. In fact, initiative was not wanted because it endangered synchronisation. Discipline was top-down and imposed.[61]

According to Lind and his co-authors, although the transition from the first generation of warfare to the second was driven mainly by technology, the principal driving force behind the appearance of the third generation of warfare was conceptual in nature. As the major technologies of third-generation warfare had been developed during the First World War, these were new tactics that broke down the stalemate of attrition created by the second-generation of warfare: 'Aware they could not prevail in a contest of materiel ... the Germans developed radically new tactics. Based on manoeuvre rather than attraction, third generation tactics were the first truly nonlinear tactics.'[62]

According to Lind, while the first two generations of modern warfare focused on 'close with and destroy' and on 'putting steel on target', the motto of the third generation was 'bypass and collapse'. Moreover, whereas second-generation warfare dispersed the order of the forces while preserving the culture of order, the third generation of warfare, based on the tactics of fast manoeuvre, made this culture obsolete:

> A Third Generation Military focuses outwards, on situation, the enemy, and the result the situation requires, not inward on process and method. Orders themselves specified the result to be achieved, but never the method. Initiative was more important than obedience. ... The Kaiserheer and the Wehrmacht could put on great parades, but in reality, they had broken with the culture of order.[63]

Similar to the first and second generations, the development of the third generation of warfare was not a sudden transformation but gradually evolved over time, with each military progressing at its own pace.[64] In 2004, for example, Lind criticised the US military for not being able to progress beyond the second generation of warfare:

> Second Generation war is relevant today because the U.S. Army and USMC learned Second Generation war from the French during and after World War I ... Aviation has replaced artillery as the source of most firepower, but otherwise (and despite the USMC's formal doctrine, which is Third Generation manoeuvre warfare), the U.S. military today is as French as white wine and cheese.[65]

To characterise the fourth generation of modern war (i.e. contemporary warfare), the authors of the 4GW theory focused on: (1) projecting the characteristics of the previous transformations on to the future nature of fourth-generation war; and (2) observing the contemporary social, political, economic and technological changes that have influenced the development of fourth-generation warfare.

On the one hand, analysing the development of the previous generations, the proponents of 4GW theory identify several characteristics of warfare that have developed and progressed from one generation to another. In his original article, Lind emphasises four central ideas that shaped previous transitions. According to Lind, each generational shift was marked by a greater dispersion of the battlefield, reduced dependence on centralised logistics, an emphasis on smaller units with higher manoeuvrability and the increasingly important need to undermine the internal power of the enemy, rather than seeking its physical destruction.⁶⁶ As Hammes concludes his observation of the first three generations:

> Each generation reached deeper into the enemy's territory in an effort to defeat him. If 4GW is a logical progression, it must reach much deeper into the enemy's forces in seeking victory ... 4GW has in fact evolved to focus deeply in the enemy's rear. It focuses on direct destruction of the enemy's political will to fight.⁶⁷

Moreover, according to Hammes, the pace of change has been accelerating:

> It took hundreds of years for the first-generation warfare to evolve. Second-generation warfare evolved and peaked in the 100 years between Waterloo and Verdun. Third-generation came to maturity in less than 25 years. Clearly, 3GW cannot be the leading edge of war over 60 years later.⁶⁸

On the other hand, in their characterisation of the fourth generation of warfare, Hammes, Lind and others also focus on the social, economic, political and technological changes that have been occurring since the end of the Second World War. Focusing on globalisation processes, the proliferation of international organisations and NGOs, economic growth, new communication and transportation technologies, the proponents of 4GW have claimed that these major societal changes have also influenced the nature of warfare, thus constituting the defining characteristics of the modern way of war. Analysing the historical fact that 'insurgency has often defeated third-generation powers since 1945', and focusing in particular on the cases of the Chinese Communist Revolution, the Vietnam War and Al-Aqsa Intifada, Hammes argues that, due to geopolitical, social, economic and technological changes, the nature of war 'shifted from an industrial age focus on the destruction of the enemy's armed forces to an information age focus on changing the minds of the enemy's political decision-makers'.⁶⁹

Combining these two understandings (the general trends of the previous intergenerational transitions and the societal changes after 1945), the advocates of 4GW theory emphasise three major characteristics of fourth-generation warfare. First, they claim that, following the trends of dispersion of forces on the battlefield, fourth-generation warfare 'seems likely to be widely dispersed and largely undefined; the distinction between war and peace will be blurred to vanishing point'.⁷⁰ Secondly, the tactical and strategic levels 'will blend as the opponent's political infrastructure and civilian society become battlefield targets'.⁷¹ And finally, as fourth-generation warfare is about the political will to fight, rather than physical fighting, 'all available networks— political, economic, social, military—[will be used] to convince the enemy's political decision-makers that their strategic goals are either unachievable or too costly for the perceived benefits'.[72] In other words, according to the advocates of 4GW, modern warfare is 'an evolved form of insurgency, rooted in the fundamental precept that superior political will, when properly employed, can defeat greater economic and military power', and it 'makes use of society's networks to carry on its fight' directly by attacking 'the minds of the enemy's decision-makers'.[73]

However, certain authors have noted a number of weaknesses to 4GW theory. According to Antulio Echevarria, the author of *Fourth-Generation War and Other Myths*, one problem with the concept is that it overstates the role of insurgency in modern conflicts:

> There is no reason to reinvent the wheel with regard to insurgencies—super or otherwise—and their various kin. A great deal of very good work has already been done … on that topic, to include the effects that globalisation and information technologies have had, are having, and are likely to have, on such movements. We do not need another label, as well as an incoherent supporting logic, to obscure what many have already made clear.[74]

The second weakness of 4GW theory, according to Echevarria, is directly related to this overly narrow focus on insurgency—not only does it completely neglect the role of national states in modern conflicts but it also offers very little in terms of solutions. While the employment of different non-military capabilities (political, economic, social, etc.) is not a new phenomenon in conflict:

> The fundamental rub, which even 4GW advocates do not address, is how to coordinate diverse kinds of power, each of which operates in a unique way and according to its own timeline, to achieve specific objectives, and to do so while avoiding at least the most egregious of unintended consequences.[75]

The theory of 'compound warfare'

The concept of compound warfare was developed in the late 1990s by military historian Thomas Huber. Huber first coined the term compound warfare in a 1997 article entitled 'Napoleon in Spain and Naples: Fortified Compound Warfare'.[76] Five years later, he republished his article in an edited volume, *Compound Warfare: That Fatal Knot*, which also contained contributions from a range of military historians who analysed different conflicts through the prism of his concept.[77]

Unlike the theories of unrestricted warfare and 4GW, which were formulated by practitioners (i.e. military or ex-military officers), the military historians who developed the concept of compound warfare did not claim it was a new phenomenon. Instead, compound warfare is merely a new conceptual framework that offers 'a new way of approaching troublesome cases where regular and irregular forces have been used synergistically' and 'the long history of warfare is replete' with such cases.[78] Like other historians who have proposed conceptual frameworks based on historical examples (such as Michael Roberts and his concept of 'Military Revolution', which was based on the example of Sweden in the sixteenth and seventeenth centuries),[79] compound warfare is a simple concept that offers very little in terms of theoretical framework:

> Compound warfare is the simultaneous use of a regular or main force and an irregular or guerrilla force against an enemy. ... The irregular force enhanced the efforts of the regular force by offering information, goods, and troops, while denying them to the enemy ... [and] the main force provides the guerrilla force with relief the enemy's presence in the locale, with training and supplies, with strategic information, and with local political leverage. ... [compound warfare is] one in which the whole is greater than the sum of the parts.[80]

In defining compound warfare, Huber highlights two fundamental characteristics: asymmetry and occupation—'Compound warfare most often occurs when all or part of a minor power's territory is occupied by an intervening major power ... [and] once the greater power's forces are distributed over the lesser power's territory, the lesser power is then in a position to conduct compound warfare.'[81]

While these two characteristics are fundamental requirements, Huber argues that compound warfare is actually a flexible phenomenon, with an enormous array of variety in this form of warfare being revealed from an analysis of historical cases. Huber claims that the concept of compound warfare generally assumes that only one side (the lesser power) may use compound warfare methods against its greater enemy (occupier). Yet, in some cases, 'both sides may use compound warfare methods ... in most historical cases of compound warfare, one uses compound warfare methods predominantly; the other side deliberately uses them to the extent it is able.'[82]

Similarly, while Huber claims that the model of compound warfare assumes only two types of force (regular and irregular), he also states that: 'Several types of mobile regional militias may fall between these two poles and may contribute importantly to the leverage of the compound warfare operator.'[83]

On the one hand, the concept suggests that the compound warfare operator ('the overall commander in a compound warfare struggle who effectively directs it') coordinates all regular and irregular forces at his disposal. On the other hand, analysing numerous historical cases, Huber concludes that: 'In the more complex reality, deliberate coordination may extend to all, some, one, or none of the military elements in play.'[84]

Finally, Huber claims that while his conceptual framework is dichotomous, assuming that a conflict is either compound warfare or not, in reality: 'One finds degrees of compound warfare. There is compound warfare proper where all the elements of compound warfare in place, and what one might call "quasi" compound warfare, where one or more elements of compound warfare are absent.'[85]

Interestingly, Huber not only criticises the simplicity of his own concept; he also defends it. According to him, the model of compound warfare is deliberately simplistic in its most basic form in order to allow for the analysis of a broad number of complex cases. In other words, Huber sought to establish a concept that would serve as a lens through which to view the history of warfare, claiming not only that it is full of cases when regular and irregular forces fought together but also that compound warfare is an especially effective form of war. And while the historical reality is extremely complex, 'the simple Compound Warfare model, it is hoped, will give analysts a place to start in coping with these complexities'.[86]

According to Huber, the most decisive and successful variety of compound warfare is what he calls 'Fortified Compound Warfare'. In fortified compound warfare, in addition to the two main components of warfare (a regular force and an irregular force), a regular force will also have access to a 'safe haven' and will be allied with a major power. This 'fortification' of the compound warfare operator refers more to an abstract sense of 'shielding from destruction' and 'strengthening' than to the construction of physical defensive positions: 'Fortification makes the difference between compound warfare, which is difficult to defeat, and fortified compound warfare, which is nearly impossible to defeat.'[87]

In other words, when a fortified compound warfare operator has a 'safe haven' for its main forces (providing the ability to withdraw the main regular force to a place inaccessible to the enemy due to geographic, technological, diplomatic, political or other factors), while also enjoying the support of a major ally, the fortified compound warfare operator 'can keep his regular force in being indefinitely ... [and] also protect and nourish the operator's guerrilla force in a similar fashion'.[88]

Huber identifies two major 'quagmire' wars, in which fortified compound warfare was used to defeat major-power adversaries: Spain from 1808 to 1814 and Vietnam from 1965 to 1973. But from the cases analysed in his edited volume, two other wars fall under the definition of this concept: the American Revolution from 1775 to 1783 and the Soviet intervention in Afghanistan from 1979 to 1989. Even if it is not clearly stated by historians, who analysed these wars through the prism of compound warfare, it seems right to argue that both cases are, in fact, examples of fortified compound warfare. In the first, Vietnam's vast territory offered the United States a 'safe haven', while France served as a major-power ally; in the second, Afghanistan's inaccessible mountain terrain offered the Mujahedeen forces protection from the Soviets, while the United States was their major ally.[89]

It is important to highlight that Huber, and other proponents of this concept, have not suggested that it is a new phenomenon. Instead, by observing and analysing the history of warfare, Huber tried to find similar patterns in which regular and irregular forces have been employed simultaneously: 'because fortified compound warfare allows operators to fight and win, in almost every historical case, with conventional force ratios that would otherwise appear to be hopelessly inferior, it is likely to be encountered often in the future.'

As such, 'Military planners need to understand the dynamics of this type of warfare before the event.'[90]

The 2005 National Defence Strategy

The last document that had significant influence on the development of the concept of hybrid warfare was the inaugural 2005 National Defence Strategy (NDS) written by Secretary of Defence Donald Rumsfeld.[91] According to Hoffman, the 2005 NDS outlined an approach required to deal with 'unconventional challenges and strategic uncertainty' and 'prepare the Department of Defence to meet 21st-century challenges'.[92] Divided into three main sections ('America's Security in the 21st Century', 'A Defence Strategy for the 21st Century' and 'Desired Capabilities and Attributes'), the NDS outlines 'an active, layered approach to the defence of the [US] nation and its interests'.[93]

While it seems that he main purpose of the NDS was an attempt to re-examine the US Department of Defense's (DoD) investment portfolio in light of the changing security environment after September 11, its analysis of the four major groups of challenges that shape the security environment in the twenty-first century was also the main value for Hoffman's conceptualisation.⁹⁴ The first group includes 'traditional challenges' that 'are most often associated with states employing armies, navies, and air forces in long-established forms of military competition'.⁹⁵ The second incorporates 'irregular challenges', posed by 'Adversaries, [who] employing irregular methods aim to erode U.S. influence, patience, and political will ... [and] take a long-term approach, attempting to impose prohibitive human, material, financial, and political costs on the United States to compel strategic retreat from a key region or course of action.'⁹⁶

The third group of challenges are called 'catastrophic challenges'. These challenges emerge from 'hostile forces [that] are seeking to acquire catastrophic capabilities, particularly weapons of mass destruction'. The fourth group comprises 'disruptive challenges' posed by potential adversaries who are seeking to develop revolutionary technologies and associated military capabilities that have a potential to fundamentally change the concepts of warfare. According to the NDS, such 'disruptive' breakthroughs might occur in biotechnology, cyber operations, space, directed-energy weapons and so on.⁹⁷

On the one hand, there is very little new in these descriptions of challenges, especially taking into consideration that the definition of irregular challenges echoes the idea of 4GW described previously. On the other, the main contribution of the NDS is its suggestion that these challenges can overlap, as 'actors proficient in one can be expected to try to reinforce their position with methods and capabilities from others'. In its analysis of this emerging capability of combined challenges, the NDS states that:

> Recent experience indicates that the most dangerous circumstances arise when we face a complex of challenges. For example, our adversaries in Iraq and Afghanistan presented both *traditional* and *irregular* challenges. Terrorist groups like al Qaeda are *irregular* threats but also actively seek *catastrophic* capabilities. North Korea at once poses *traditional*, *irregular*, and *catastrophic* challenges.

Moreover, concluding this statement, the NDS claims that 'in the future, the most capable opponents may seek to combine truly disruptive capacity with traditional, irregular, or catastrophic forms of warfare.'⁹⁸

It is important to emphasise two main aspects of the NDS. On the one hand, focusing on the highest strategic level and written for the specific purposes of preservation and enhancing America's competitive advantages in light of the new security environment, the NDS sets a very traditional framework of different emerging challenges. On the other, it acknowledges that the division between these challenges is becoming blurred and that this merger will produce America's most capable adversaries. In other words, without delving into the details of this fusion, as this is not its purpose, the NDS suggests that tomorrow's challenges will not present very distinct and separate threats, but a dangerous mixture of them.

2

THE BIRTH OF 'HYBRID WARFARE'

The previous chapter discussed the conceptual debate over the nature of warfare that preceded the development of the hybrid warfare concept. This chapter now turns to the characteristics of this latter concept as it was initially formulated by Hoffman. It begins with an analysis of Hoffman's works, as well as those of other US military thinkers, before examining the concept of hybrid warfare in the context of US strategic culture and how the concept differs from the theories discussed in the previous chapter. The chapter aims to provide a more complete overview of the original concept, its characteristics in US military discourse and the reasons behind its success.

Although hybrid warfare refers to 'the blurring and blending of previously separate characterisations of different modes of warfare',[1] it can be argued that Hoffman's concept was itself based on a hybrid of different strategic ideas and theories that were developed between the late 1990s and the early 2000s. In addition to the four sources discussed in the previous chapter that influenced Hoffman in his development of hybrid warfare—namely unrestricted warfare, 4GW, compound warfare and the NDS—he also acknowledges the work of other scholars, strategists and military thinkers: 'Many other analysts have captured these trends, with Russian, Australian, and American authors talking about "multi-modal" and "multi-variant" forms of war.'[2]

Specifically, Hoffman highlights the importance of the works of Michael Evans, Stephen Blank, Colin Gray, John Arquilla, Bruce Hoffman and John Robb in the formulation of his concept.[3]

Hoffman consequently bases his theory of hybrid warfare on a combination of observations made by other thinkers on modern warfare:

> From the 4GW school, it [hybrid warfare] uses the concept of the blurring nature of conflict and the loss of the State's monopoly of violence. The concepts of omni-dimensionality and combinations were crucial ideas adopted from Chinese analysts. From John Arquilla and T.X. Hammes we took in the power of networks. From the proponents of Compound Wars, the concept absorbs the synergistic benefit of mixing conventional and unconventional capabilities, but at lower and more integrated levels. From the Australian experts, we have accepted the growing complexity and disaggregated nature of the operational environment, as well as the opportunistic nature of future adversaries.[4]

This intellectual synergy between different schools of strategic thinking[5] allows Hoffman to articulate two definitive and conceptually interconnected terms. The first of these is hybrid war, which: 'Incorporates a range of different modes of warfare including conventional capabilities, irregular tactics and formations, terrorist acts including indiscriminate violence and coercion, and criminal disorder.'[6]

The second is 'hybrid threat': 'Any adversary that simultaneously and adaptively employs a fused mix of conventional weapons, irregular tactics, terrorism, and criminal behaviour in the battlespace to obtain its political objectives.'[7]

Although Hoffman borrows certain ideas from the theory of unrestricted warfare—such as general conclusions about combined threats and the blended nature of modern conflict, as well as more specific observations about a threat presented by a possible 'cocktail' of high-tech capabilities with terrorism and cyber-warfare directed against financial targets[8]—Hoffman's concept of hybrid warfare is a much more practical concept. Unlike the concept of unrestricted warfare, which tries to grasp the whole picture, the benefits of which were greatest at the strategic level, Hoffman's concept of hybrid threats focuses chiefly on tactical and operational military activities 'directed and coordinated within the main battlespace to achieve synergistic effects'.[9] In other words, the concept of hybrid warfare was intended to be of greater use to military practitioners than grand strategists, and it sought to satisfy the understanding of 'the kind of threats the Marine Corps would face in the future'[10] rather than being used as a strategic model of future human conflict.

Hybrid warfare is also heavily based on the theory of compound warfare and its concept of synergetic benefits from a simultaneous employment of regular and irregular forces. However, unlike the concept of compound warfare, which draws a distinction between regular and irregular forces on the tactical and operational levels, as well as between the forces' capabilities and contributions to the strategic aim of the campaign, hybrid warfare is conducted by opponents that 'seek victory by the fusion of irregular tactics and the most lethal means available'. Acknowledging the contribution of the proponents of compound warfare to his own concept, Hoffman does not claim that there is any novelty in the combination of regular and irregular components. Instead, he claims that while 'in most conflicts, these components occurred in different theatres or in distinctively different formations', in hybrid wars 'these forces blurred into the same force in the same battlespace'.[11]

By adding terrorism and criminality to the irregular forces emphasised in the concept of compound warfare, Hoffman also brings in a number of ideas formulated by the proponents of 4GW, such as society's networks, the dispersed and largely undefined nature of conflict, as well as the loss of the state's monopoly on violence. Thus the idea outlined in the concept of 4GW of defeating the political will of decision-makers is echoed in Hoffman's suggestions that:

> operationally integrated and tactically fused, the irregular component of the force attempts ... to provoke overreactions or extend the costs of security for the defender ... The goal [in hybrid war] may include protracted conflicts with a greatly defused set of force capabilities to wear down resistance, or the actual defeat of a conventionally-oriented government.[12]

Yet unlike 4GW, which focuses exclusively on insurgency and claims that fourth-generation warfare can be used only by sub-state actors, Hoffman states that hybrid warfare can be conducted by states and a variety of non-state actors, as well as by their different combinations:[13]

> Hybrid Threats blend the lethality of state conflict with the fanatical and protracted fervour of irregular warfare. In such conflicts, future adversaries (states, state-sponsored groups, or self-funded actors) exploit access to modern military capabilities including encrypted command systems, man-portable surface-to-air missiles, and other modern lethal systems, as well as promote protracted insurgencies that employ ambushes, improvised explosive devices, and assassinations.[14]

The narratives contained in the 2005 NDS can also be traced within the definition of hybrid threats. However, while Hoffman states that he used the four types of challenges to formulate his idea of hybrid war, he did not adopt these ideas without criticism.[15] Thus he praises the NDS's suggestion that the most complex threats will seek synergies by combining different challenges and that the threat categories could overlap, thereby creating the most capable opponents and adversaries.[16] Yet he also replaces catastrophic threats with terrorism, 'which could be catastrophic in impact', and disruptive challenges with criminality, which is derived from 'disruptive social behaviour'.[17]

According to Hoffman, the concept of hybrid warfare:

- Describes the evolving character of conflict better than counterinsurgency
- Challenges current 'conventional' thinking and the binary intellectual bins that frame the debate
- Highlights the true granularity or breadth of spectrum of human conflict
- Raises awareness of potential risks and opportunity costs presented by the various options in the ongoing threat/force posture debate.[18]

However, the main contribution of Hoffman's conceptualisation stems from its practicality, especially to military decision-makers at the operational level. Unlike the proponents of compound warfare, who limited themselves to historical observations, Hoffman pays much more attention to the conceptual foundations of his theory. Unlike the Chinese officers and their attempt to create an all-encompassing analysis of the changing nature of the human conflict, Hoffman's analysis is restricted to the operational level, making hybrid warfare comprehensive enough to incorporate the big ideas of unrestricted warfare, but also simple enough to be useful to practitioners. Finally, and most importantly, unlike the advocates of 4GW, who offer very little in terms of solutions, Hoffman lays down an explicit list of implications for the US military in terms of force-planning, doctrine and organisation, training and education.[19] In other words, unlike his predecessors, Hoffman not only tries to delineate and define the nature of his theory; he also seek to 'explain its relevance to today's ongoing force posture debate',[20] offering not only a new concept based on the analysis of historical cases but also a detailed recipe for the required adaptations to the new realities.

When discussing the genesis of the theory of hybrid warfare, it is impossible to avoid the main example that was used by Hoffman to demonstrate the relevance of his conceptualisation—the Second Lebanon War in 2006, between Israel and Hezbollah. In his analysis of the different conflicts of the twentieth century, ranging from the Irish insurgency of 1919–20, through the Mujahedeen in Afghanistan in the 1980s and up to the Balkans experience of the post-Yugoslavia era and the fight of Chechen rebels against the Russians in the 1990s, Hoffman concludes that 'in all cases, we found conventional and irregular tactics, terrorism, as well as criminal activity'. Yet these conflicts lacked 'the multi-dimensionality' and 'operational integration' that he claims will characterise the hybrid wars of tomorrow. Although he labels these conflicts as 'the earliest prototypes' of hybrid war, the true and most comprehensive prototype in his eyes was the Second Lebanon War, and specifically Hezbollah, which he describes as 'the clearest example of a modern Hybrid challenger'.[21]

Hoffman's four main observations of the Second Lebanon War lead him to conclude that 'Hezbollah is [a] representative of the rising hybrid threat.'[22] The first observation is that Hezbollah represented a true fusion between regular and irregular forces. Hezbollah's forces were highly disciplined, well trained and properly equipped units that resembled a regular army more than had been the case in the other insurgencies fighting against the IDF in the West Bank or the Gaza Strip. Yet, at the same time, these forces were organised in decentralised cells that employed adaptive guerrilla tactics in both urban centres and mountainous areas of southern Lebanon. The second characteristic that led Hoffman to refer to Hezbollah as a hybrid actor was the organisation's simultaneous use of myriad technologies on the tactical, operational and strategic levels. While Hezbollah's arsenal was composed of many sophisticated anti-tank weapons, which it used to target Israeli tanks and other armoured vehicles, and a vast number of short-range standardised rockets, it also included armed UAVs, anti-ship cruise missiles and sophisticated technologies for gathering signal intelligence.[23]

The third aspect of Hezbollah's actions that attracted Hoffman's attention was the organisation's successful use of military actions and information operations on the strategic and operational levels. While, in pure military terms, Hezbollah's achievements were dubious, the organisation effectively exploited information warfare to magnify the political effects of its limited tactical successes. Describing the strategic battle of perceptions during this conflict, Hoffman concludes that despite significant losses to Hezbollah's 'most dedicated and trained portion of the militia's ground force ... The IDF's credibility has been weakened and Hezbollah arguably came out of the conflict stronger in ideological appeal.'[24]

The final characteristic of Hezbollah's fighting that Hoffman uses as a prototype for his concept of hybrid war was its use of these combinations simultaneously, by one organisation in a single battlespace over a very short period of time, which provided an advantage against a much stronger (in traditional military terms) Western-style military. In other words, when militia forces and highly-trained fighters on the same battlefield are combined with the use of low-tech rockets and high-tech UAVs, cruise missiles and advance surveillance equipment, bolstered by an effective form of information warfare that magnifies the effects achieved on the ground, it does not fall under any existing definitions of the different modes of warfare, but represents a certain hybrid of them.

It is important to note that Hoffman stops short of suggesting that the kind of hybrid warfare practised by Hezbollah in 2006 was a definitive demonstration of what hybrid wars would look like. The opposite is true. He acknowledged that Hezbollah's decision to combine different capabilities and tactics was the result of specific conditions and circumstances, such as the organisation's ability to prepare terrain for a fight against a single, well-studied and recognised enemy, or to benefit from relations with Iran and Syria, two countries that supplied Hezbollah with money, arms and training expertise.[25] Therefore, according to Hoffman, Hezbollah's form of hybrid warfare is just one example of a vast number of possible hybrids, or, using the language of unrestricted warfare, it is simply one specific 'cocktail' from a large menu of hybrid threats. For example, to demonstrate the versatility of his concept, Hoffman analysed the experience of the US Navy against Iran in the Persian Gulf in the late 1980s, thus suggesting that his concept of hybrid warfare is not limited to land warfare and is equally relevant to understanding emerging threats in the maritime realm.[26]

Hoffman's conceptualisation of hybrid warfare offers two major advantages over the theories and concepts discussed previously. First, by interweaving different ideas from the strategic thinking of the late 1990s and early 2000s, Hoffman puts forth a synergetic theory that not only encompasses all their advantages but also bypasses their limitations. The abstract, all-embracing, and to some degree philosophical, approach of unrestricted warfare was grounded by military practicability of hybrid war. The limited theoretical conceptualisation of compound warfare, an idea formulated by military historians who were primarily interested in historical cases rather than conceptual patterns, is compensated for in Hoffman's theory by a more practical, comprehensive focus. And the overly narrow focus on insurgency in 4GW is replaced by the highly flexible idea of hybrid threats in Hoffman's work.

The second advantage of hybrid warfare over the theories and concepts discussed thus far is that Hoffman situates his ideas in the context of an ongoing, policy-relevant debate. Due to its engagement with official publications and provision of specific policy recommendations and solutions, Hoffman's theory rapidly gained in popularity in US military circles.

THE BIRTH OF 'HYBRID WARFARE'

The rise of hybrid warfare

Although Hoffman was not the first to notice that different modes of warfare are in the process of amalgamating with one another, and though his use of hybrid warfare was not the first concept to deal with this phenomenon, the concept has undoubtedly been very popular, particularly in the US military. There are four reasons for this success. The first, and most obvious, is that hybrid warfare, unlike previous theories, offers an operationally applicable concept that directly addresses the evolving character of modern conflict from an explicitly military perspective. In other words, the military found it more intuitive given that it was designed to be used at the tactical–operational level, rather than being a grand strategy that involved politics, economics and culture (as, for example, with unrestricted warfare or 4GW). Another reason for the success of the concept was probably its author, and his distinguished record as a Marine Corps reserve lieutenant colonel, which includes service on two national commissions and three defence science boards; he has also held a position as a research fellow at the Centre for Emerging Threats and Opportunities at the Marine Corps Combat Development Command, as well as writing several books and hundreds of articles.[27] He was just the right man, at the right place and and at the right time, to cast his ideas out.

This leads to the next reason for the success of hybrid warfare as a concept. As an active author and influential scholar, Hoffman ensured that his concept would reach every corner of the professional military debate by publishing articles and manuscripts in a wide range of outlets, including *Naval Institute Proceeding*; *Armed Forces Journal International*; *Defense News*; *Marine Corps Gazette*; *Small Wars Journal*; *Joint Force Quarterly*; *Strategic Forum*; *Orbis*, and many others,[28] including his manuscript, *Hybrid Warfare*, published by the Potomac Institute for Policy Studies and described as 'a masterpiece of enlightened thinking' by the institute's chairman.[29] The final explanation for the success of Hoffman's conceptualisation is its use of the Second Lebanon War as an example. By choosing Hezbollah as a model for his conceptualisation of hybrid threat, Hoffman's theory has an enormous advantage over everything that was written before. The Israeli experience in Lebanon was a 'hot potato' among US military scholars for several years after the end of the war,[30] allowing Hoffman to slip his concept into the ongoing debate about the lessons that the US military should learn from this conflict.

Thus the concept of hybrid warfare has become extremely popular within US military circles, generating an enormous amount of literature on the subject. Although it is not possible to list the numerous works here, it is worth mentioning some of the most significant. The first wave of military thinkers writing about hybrid warfare in the late 2000s included Colonel Steven Williamson, who traced the conceptual links between the 4GW theory and the concept of hybrid warfare;[31] Colonel Margaret Bond, who applied the ideas of hybrid warfare to stability operations in failing states;[32] Lieutenant Colonel Daniel Lasica, who dealt with the difficult question of victory in hybrid wars;[33] Major Larry Jordan, who analysed US Army doctrine through the prism of hybrid warfare, making several constructive recommendations for the military decision-makers;[34] and Major Sean McWilliams, who produced a useful analysis of the 1976–89 war in South Africa as an example of hybrid warfare, thus suggesting that, while the concept itself might be novel, the phenomenon was not new at all.[35]

The second wave of publications sought to provide deeper insights into the concept of hybrid threats[36] and the implications of this concept for the rules of engagement.[37] The two most important publications, however, appeared in 2010, confirming the centrality of the concept of hybrid warfare in the professional military debate. The first was 'Military Capabilities for Hybrid War,' a report published by the highly influential RAND Corporation, which described Israel's adversaries in Lebanon in 2006 and Gaza in 2008 as hybrid threats, or, in other words, 'neither an irregular opponent nor a state actor'.[38] The second publication was a briefing paper on hybrid warfare published by the US Government Accountability Office and presented to the Subcommittee on Terrorism, Unconventional Threats and Capabilities in the US House of Representatives.[39] While adding nothing new to the theory, the implication of these documents for the success of the concept of hybrid warfare was fundamental as they drew the attention of the American political establishment to the ongoing debate within academic and professionial military circiles.

In the following years, interest in Hoffman's concept of hybrid warfare has continued to grow, with analysts focusing on the additional implications of the theory, such as its ethical challenges[40] and the increasing use of information warfare.[41] However, the most comprehensive analysis of the subject is contained in a book from 2013 by two junior officers, Major Timothy McCulloh and Major Richard Johnson,[42] entitled *Hybrid Warfare*. The book not only focuses on a series of historical examples, ranging from the Second World War to Operation Iraqi Freedom, but it also offers an overall assessment of the concept and its relevance to future conflicts; as such, it deserves a more detailed treatment.

THE BIRTH OF 'HYBRID WARFARE'

McCulloh and Johnson's analysis is vital for understanding the developments in US military thought on hybrid warfare, as it contributed both to the development of the concept and its implications for use in practical situations. The first part of the book starts with a comprehensive analysis of the conceptual debate over the theory of hybrid warfare. Following a thorough historical examination of Hezbollah's tactics in 2006 and the Soviet partisan network on the Eastern Front during the Second World War, McCulloh concludes that: 'Historically, the hybrid formation process has resulted in several commonalities in terms of composition and effects, which in turn can be generalised into seven principles to describe hybrid war in its totality.'[43]

The first principle of hybrid warfare, as formulated by McCulloh, is that the composition, capabilities and effects of a hybrid force are unique to 'the temporal, geographic, socio-cultural, and historical settings in which the given conflict takes place'. In the case of Hezbollah, this principle was reflected in the weak central government in Lebanon, which allowed the organisation to maintain its militant status; a large Muslim Shi'a population that served as a solid base of support; Iran and Syria as external sponsors; and, finally, Hezbollah's ideology, which allowed it to garner sympathy and support from the Lebanese population. Regarding the Soviet partisan network, McCulloh emphasises the harsh terrain of the eastern Russian steppes; the Russian experiences during the First World War and the Civil War that followed the Bolshevik Revolution of 1917; and Communist ideology. All of these factors served as a preconditioned context that allowed for the militarisation of the Soviet people, enabling the recruitment of large portions of the population to form the hybrid partisan network.[44]

The second principle is the existence of 'a specific ideology within the hybrid force that creates an internal narrative to the organisation'. In the case of Hezbollah, McCulloh emphasises the importance of the ideology stemming from the Iranian Islamic Revolution for maintaining the status of the organisation as an anti-Israeli militia and as a protector of the Shi'a in Lebanon; in the case of the Soviet partisan network, it was the ideology of Communism that 'made both loyal personnel and physical resources readily available to any entity which supported the state's desires—specifically to both the Red Army and the Partisan Network.'[45]

The third principle is the idea that a perceived existential threat from a potential adversary 'drives the hybrid force to abandon conventional military wisdom to achieve long-term survival'. Following McCulloh's examinations of historical cases, this principle claims that, when facing an existential threat, a hybrid force will seek 'any method possible to defend itself ... including the unconventional, terrorist, and criminal activities that support the organisation'. This was the case with Hezbollah, which had already experienced Israeli aggression and was being described in harsh rhetoric in Israeli political statements; this was also the case with the Soviet partisans, who were experiencing German aggression, which was in turn fuelled by Nazi ideology. The fourth principle is the asymmetry between hybrid forces and their potential adversaries (i.e. when a hybrid force is significantly smaller in terms of conventional military capabilities, it will 'seek a way to offset this apparent advantage [of its adversary]').[46]

The fifth principle of hybrid warfare, according to McCulloh, is that 'a hybrid force contains both conventional and unconventional elements' in terms of technologies, weapons and tactics. The sixth states that 'hybrid organisations rely on inherently defensive type operations'. While these operations might include offensive elements, their primary aim will be of a defensive nature. The seventh, and final principle, is based on the idea that hybrid organisations impose a war of attrition on their adversaries in both the physical and cognitive domains.[47]

McCulloh's conceptualisation offers a useful set of parameters to characterise hybrid warfare, buttressed by a comprehensive analysis of historical cases. However, it also suffers from two main disadvantages. First, the distinction between his seven principles is unclear—they often tend to overlap with similar narratives, such as ideology and asymmetry, which are also used by McCulloh to characterise the different principles. Secondly, and more importantly, McCulloh's conceptualisation does not clarify whether all seven principles are required to define an adversary as hybrid, or whether some lesser combination of them would suffice.

These disadvantages notwithstanding, McCulloh offers an additional and useful conceptual layer that was absent in Hoffman's original conceptualisation, thereby ensuring that the concept was sufficiently abstract to be relevant for a variety of cases, yet explicit enough to grasp the phenomenon with accuracy.

The second part of the book, written by McCulloh's co-author Richard Johnson, focuses on the US experience in Vietnam, as well as the US intervention in Iraq as part of Operation Iraqi Freedom, in order to explore the implications of hybrid warfare for the operational art of warfare. Unlike McCulloh, who focuses more on the characterisation of hybrid forces, Johnson emphasises the requisite operational approaches to defeat hybrid threats. In his analysis of the US military's varying levels of success in fighting hybrid adversaries in Vietnam and Iraq, Johnson suggests three main imperatives for the operational art in hybrid warfare.[48] First, he claims that the main aim of the operational art is to break the hybrid adversary's inherent logic, rather than focusing on the physical destruction of its forces: '[It] must cognitively disrupt the hybrid threat's logic in the forms of warfare it employs, rather than focusing on physical methods to counter the hybrid threat's means and capabilities.'[49]

Secondly, a successful operational approach must fuse tactical success with strategic aims within the same context that gave rise to the hybrid threat. Analysing the transformation of the operational art during Operation Iraqi Freedom, Johnson concludes that a successful operational approach must not only address the underlining accelerants that drive hybrid threat activity; it must also arrange tactical actions in the same cultural, social, ideological, political and other frames of reference that maintain the success of the hybrid adversary.[50]

The third imperative for a successful operational approach in a hybrid war, according to Johnson, is that those fighting it 'must avoid prescriptive or uniform measures across time and space'. As the environmental context of a hybrid threat constitutes a complex system with specific characteristics that constantly change across time and space, 'the operational approach cannot simply give commanders an appreciation for the complexity of the problem, while dogmatically refusing to bound it'. Instead, the operational approach has to balance between general guidance and doctrinal directives and the freedom of tactical commanders to develop tailored solutions and internal measures of success.[51]

McCulloh's and Johnson's analyses have thus made a significant contribution to the debate over the nature of hybrid warfare by looking at different historical cases through the prism of hybridity and strengthening the conceptual framework underlying the theory. Yet one of the most important aspects of their work is its practical nature. The book engages with relevant doctrinal publications, providing a set of practical implications required not only for a better understanding of the nature of future adversaries but also the different ways to defeat them.

This leads to an additional reason for the success of hybrid warfare as a concept that allows for a greater understanding of the blending of the different modes of warfare: namely US military culture. There is no need to repeat here what has already been said elsewhere about US military culture or the American way of war, both of which have been widely discussed in the literature;[52] however, it is important to highlight the aspects of US military culture that have been most relevant to the success of hybrid warfare as a concept.

Proponents of the cultural approach to military affairs have highlighted the influence of cultural narratives on the way that nation states conduct conflicts.[53] The US case is not an exception, as 'geography, history, cultural heritage, and long-held political objectives [have] influenced the American conduct of war'.[54] In its analysis of US military culture, the existing research repeatedly points to three principal characteristics: the aggressive and decisive employment of military personnel and firepower to conclude a given conflict as quickly as possible; heavy reliance on state-of-art technology; and the apolitical nature of the traditional US military mind-set, which gives 'little regard to the non-military consequences of what they were doing'.[55] Although some of the research has suggested that these cultural characteristics have actually created significant difficulties for the US military when dealing with non-traditional adversaries,[56] they also help to explain the success of hybrid warfare as a concept.

Unlike the classic Clausewitzian observation that 'war is merely the continuation of policy by other means',[57] many scholars have claimed that the US way of war has traditionally been built on the assumption that 'war [is] not a continuation of political intercourse, but a symptom of its failure'.[58] The US Army's textbook on strategy, published in 1936, simply states that: 'Politics and strategy are radically and fundamentally things apart. Strategy begins where politics end. All that soldiers ask is that once the policy is settled, strategy and command shall be regarded as being in a sphere apart from politics.'[59]

In other words, culturally, the US military mind-set traditionally tends to draw a distinction between military actions and other dimensions of human conflict, such as politics and economics. This tendency is obviously not unique to the US military. However, it can be argued that US military culture has consistently failed to grasp the interconnection between all of these dimensions. Moreover, as Americans 'have approached warfare as a regrettable occasional evil that has to be concluded as decisively and rapidly as possible',[60] it is not surprising that scholars have identified a tendency within the US military tradition to value practice more than theory, and to neglect the importance of the kind of professional military education required for preparation for a future conflict.[61] Traditionally, Americans have believed in their technological and logistical superiority,[62] taking it for granted that it will enable them to concentrate enough forces and firepower to achieve what they consider as the main objective of war—'the complete overthrow of the enemy, [and] the destruction of his military'.[63] Consequently, it is hardly surprising that US military culture has consistently failed to embrace a comprehensive understanding of strategy that extends the dimension of pure military actions: it has simply never had to.[64]

Furthermore, as many scholars have emphasised, US military culture tends to view technology as a lodestar of military success, as 'no nation in recent history has placed greater emphasis upon the role of technology in planning and waging war than the United States'.[65] Deeply rooted in the US character, this fascination with technology and science has often been used as a panacea in national security affairs, with the US military spending 'hundreds of billions to replace the most advanced system on the planet with another system only slightly more advanced'.[66] In the US military mind-set, this technical romanticism in military affairs has shaped not only the understanding that warfare is a rivalry between technologies but also a pragmatic approach towards complicated issues, viewing them as problems that require engineering solutions.[67] The traditional orientation towards quick actions and results, an inherent tendency to grasp the complex realities of warfare from a purely military point of reference, and a bias towards a techno-centric understanding of warfare have encouraged the US military 'to fight highly organised, systematic, material, and technology based wars'.[68] Thus, this predisposition to simplify warfare to a problem that can be solved quickly by the military with the right technological tools formed an inherent preference for the US military for the more feasible conceptualisation of warfare, at the expense of a more all-embracing and complete comprehension of the nature of the human conflicts. In other words, culturally, the US military mind-set values concepts that are workable and grounded in what the military is capable of doing, rather than focusing on abstract academic theories. Hoffman's concept of hybrid warfare dovetails with these cultural preferences.

Written by the US military and for the US military, the concept of hybrid warfare addressed the two basic demands of US military culture—a purely military orientation and conceptual practicality. First, Hoffman's concept transferred the problem of a complex and amorphous adversary that did not fall into previously established boxes to the purely military realm, thus creating a new conceptual framework that the military is able to understand. Secondly, Hoffman and the other proponents of hybrid warfare have consistently sought to emphasise the practical nature of the theory, not only debating what hybrid wars and threats are but also what can be done at the practical level to face such threats successfully. Rather than the abstract and 'omnidirectional' nature of unrestricted warfare, hybrid warfare offers a similarly complex but more feasible concept, one that can be solved by military means. In place of the historical, and under-conceptualised, observations associated with the proponents of compound warfare, the concept of hybrid warfare provides a detailed conceptual vocabulary with which military practitioners are able to work. And finally, instead of the almost apocalyptic observations of 4GW without solid policy recommendations, the concept of hybrid threats not only identifies the problem but also suggests practical solutions that have implications for the evolution of military doctrine.

To summarise the origins and rise of hybrid warfare as a conceptual explanation of the early twenty-first-century security environment and the threats it creates, it is important to focus on its three main aspects as a theory of warfare. The first is its conceptual novelty. On the one hand, from the analysis presented above, hybrid warfare is merely a new conceptualisation of an already existent phenomenon. The historical analyses show that it has been an integral part of warfare since antiquity.[69] The only aspect that is new is that certain factors, such as contemporary military technologies and the means of communication (the internet, cellular phones, social networks, etc.), have expedited the blend between regular and irregular forces, operational concepts and organisational methods. Hoffman himself acknowledges that his concept does not describe a novel phenomenon but rather an enhanced and simultaneous fusion of different capabilities, tactics and methods that were previously separated in space and time.[70] Yet the novelty of hybrid warfare at the time of its conceptualisation was that, unlike previous theories, it offered a concept that was operationally feasible, generating practical solutions that could be applied by military professionals.

This leads to the second aspect—the instrumental approach of hybrid warfare in dealing with neither regular, nor irregular, threats. Discussing the problem of emerging hybrid threats, Hoffman and others address the problem in pure military terms, isolating it at the operational level of military decision-makers. In other words, hybrid warfare and hybrid threats describe problems that could be solved by the military once it has a suitable doctrine, is trained and equipped accordingly and enjoys the required support of the political leadership. From the very beginning, this theory was intended to improve the practical performance of military units on the battlefield, engaging with relevant official publications and providing specific recommendations and solutions. In Hoffman's words:

> [If we] simply gain a better understanding of the large grey space between our idealized bins and pristine Western categorizations, we will have made progress. If we educate ourselves about how to better prepare for that messy grey phenom-

THE BIRTH OF 'HYBRID WARFARE'

enon and avoid the Groznys, Mogadishus and Bint-Jbeils of our future, we will have taken great strides forward.[71]

The final aspect is the success of hybrid warfare as a theory in military and academic discourse. On the one hand, Hoffman's ideas have successfully captured the imagination of Western military thinkers, scholars and strategists. While the theory has proven particularly popular among the US military establishment, it has also succeeded in attracting international recognition with many non-American scholars joint the conceptual discourse.[72]

On the other, it is important to note that, despite the efforts of this theory's proponents, even the US military establishment has been reluctant to integrate hybrid warfare theory into its doctrinal publications. Though certain ideas stemming from the concept have found their way into some US Army publications,[73] it continues to be treated more as 'an alternative concept about the ever-evolving character of modern conflict'[74] than an officially and widely adopted concept. Indeed, even the US Department of Defense's *Dictionary of Military and Associated Terms* does not list any definitions for hybrid warfare or hybrid threats.[75]

PART 2

THE RISE OF *GIBRIDNAYA VOYNA*: THE RUSSIAN THEORY OF HYBRID WARFARE

3

READING EVGENY MESSNER

THE THEORY OF 'SUBVERSION-WAR' (*MYATEZHEVOYNA*)

Evgeny Messner was an Imperial Russian émigré officer whose books were prohibited in the USSR due to his strong anti-Communist views. After the Cold War, however, his works have become increasingly popular, taking a more central place within the Russian school of military thinking.[1] As Messner's works have remained largely unknown to the Western reader, it is important to introduce him and his career as it significantly shaped his views on the phenomenon of war in the twentieth century.

Evgeny Messner: a man of practice and theory

Evgeny Eduardovich Messner was born on 3 (16 in the Gregorian calendar) September 1891 in the Kherson governorate in the southern Ukrainian region of the Russian Empire. His great grandfather was a Württemberg noble who migrated to Russia during the reign of Catherine the Great. While his father was a Lutheran and his mother Catholic, his parents raised Evgeny and his younger brother Victor according to the Russian Orthodox tradition, 'celebrating at home all Orthodox holidays and giving their children names that would allow them to convert to Orthodoxy without changing their names'— which they did indeed do before getting married.[2]

After leaving school in Odessa, Evgeny was admitted to the faculty of physics and mathematics at the Imperial Novorossiya University.[3] However, his studies did not last long, as Evgeny left after a year to become a military officer.⁴ Later in his life, Messner recalled:

> I decided to become an officer from a very young age. Destiny [however] opposed it: after the end of the gymnasium, for two consecutive years I could not enrol into the Mikhailovsky Artillery School due to health problems (first typhoid fever, and in the following year a serious football injury). To make up for the lost time, I followed the advice of my father's friend, Colonel Mende, a military prosecutor, to choose a fast, but incredibly difficult path to the officer ranks: to take the School's [final] examinations as an external student.⁵

Following his decision, Messner moved to St Petersburg, where he found a school that prepared prospective students for entry to military schools. For an 'astronomic sum of money', the school arranged for him to be tutored by a special group of officer–academics (officers who had graduated from military academies) who prepared him for taking the final examinations of the Mikhailovsky Artillery Academy. Messner proved to be a very talented student, finishing the programme in six months rather than a year, as was recommended by his instructors.⁶ In April 1912, Messner successfully passed the final examinations at the academy, one of the finest military educational institutions in the Russian Empire, with a first-class degree. For an external student, such high results in the final examinations were very rare and considered an exceptional achievement. In later years, when recalling his success, Messner wrote: 'This success I owe to my working capabilities ... and my exceptional memory, and thirdly, or maybe firstly, my desire to be an officer and follow in the footsteps of my beloved uncle Alexander (an officer, and later General Staff General).'⁷

Messner's father—an architect at the Offices of the Institutions of Empress Maria (the governmental office of charity in Imperial Russia), with responsibility for building and renovating orphanages in Odessa—was very supportive of his son's desire to pursue a military career, yet it seems that it was his uncle, Alexander Yakovlevich Messner, who was the role model for young Evgeny. This family connection, to a General Staff colonel, helped *Podporuchik* (Second Lieutenant) Messner to be assigned on 3 October 1912 to the 5th Battery of the 15th Artillery Brigade, stationed in his hometown Odessa.⁸

Messner proved himself to be a highly energetic and talented officer, and he quickly started to climb up the military hierarchy. On 13 March 1913, he was transferred to the training unit of the brigade as an instructor—which was very unusual for a young *podporuchik* with only six months' service, as the position was traditionally reserved for a *stabs-kapitan* (senior lieutenant) or, at least, an experienced *poruchik* (lieutenant). Less than a year later, on 30 January 1914, Messner was again promoted to a position of acting (*ispolnyayushchiy dolzhnosti*—as this position was much higher than his junior rank) adjutant of Colonel Sergey Lukashevich, the commander of the 2nd Squadron of the brigade.⁹

During the First World War, Messner continued to climb up the ranks, proving himself a talented and courageous officer. On 15 April 1915, he was raised to the rank of *poruchik*. In August of the same year, he was evacuated to Odessa due to a serious illness, where he married Lyudmila Emmanuilovna Kalinina, to whom he had been engaged since 1913. Lyudmila was the daughter of retired General Staff Lieutenant General Immanuel Kalnin, who had originally opposed their marriage before his future son-in-law had completed the General Staff courses. However, due to the war, he changed his mind, and the couple got married.¹⁰

On 23 October 1916, already a *stabs-kapitan*, Messner was sent to complete the academic courses at the Imperial Nicholas Military Academy (the General Staff academy), eventually finishing in the top ten of his class.¹¹ In early February 1917, Messner went back to the front and was soon appointed to the position of acting chief of staff of division. At the front, he found a disintegrating army that was infected by revolutionary ideas and a general unwillingness to continue fighting. As a professional officer, Messner was deeply frustrated by the populist, 'revolutionary' narratives that had taken over the hearts and minds of many in the military.¹² Despite this desperate situation, Messner continued to perform his duties until the early spring of 1918 when his division was demobilised by the newly established Soviet government. In March 1918, Messner arrived in Odessa with the remnants of his division. Messner reflected on his experience of the First World War as follows: '[I] withstood more than 1,000 days of hard battles, was wounded and contused seven times, received 12 medals, including the Cross of Saint George, and rose from a modest post of an Adjutant of artillery squadron to the position of the Chief of Staff of division.'¹³

Messner joined the White Movement almost immediately after his return to Odessa in the late spring of 1918. He served in different positions during the Civil War, taking an active role in fighting the Bolshevik forces, most notably as the last chief of staff of the Kornilov Division of General Wrangel's army. In November 1920, General Staff Colonel Evgeny Messner left Russia from Crimea with the last ships that evacuated the defeated White forces.¹⁴

After leaving Russia, Messner moved to Belgrade, where he took an active part in the social and military–academic life of the Russian émigré community. His works on military theory and tactics were published in different military periodicals by the Russian communities abroad, such as *Voyennyy Sbornik* (Military collection), *Russkiy Voyennyy Vestnik* (Russian military bulletin), and *Vestnik Voyennykh Znaniy* (Bulletin of military knowledge), among many others. In 1931, when a branch of the school of Higher Military Science Courses of General Nikolai Golovin was opened in Belgrade, Messner was invited to take up a post as a lecturer. During this period, he proved himself to be a prominent anti-Communist and took an active role in the political and social life of the Russian All-Military Union (an organisation founded by General Wrangel to unite the Russian White Movement veterans).[15]

After the start of the Second World War, Messner continued to lecture at the Higher Military Courses in Belgrade, preparing officers for the Russian Corps (an armed force composed of anti-Communist Russian émigrés in the territory under the control of the German Military Command in Serbia). Until the spring of 1945, Messner served in the military-propaganda department of the Wehrmacht 'South East Army', where he led the Russian section and was an active supporter of the establishment of the Russian Liberation Army (the Vlasov Army—a group of predominantly Russian forces that fought under German command). In March 1945, Messner took the position of the head of the propaganda department in the 1st Russian National Army, established under the command of Russian émigré General Boris Alexeyevich Smyslovsky-Holmston. After the army's capitulation in Lichtenstein in May 1945, and Liechtenstein's refusal to expatriate the members of the army to the Soviet Union, Messner, together with his wife, moved to Argentina in the autumn of 1947.[16]

In Argentina, Messner continued his earlier work as a journalist, publisher, writer and military theorist. One of his most prominent achievements was the establishment of the South American branch of the Institute for the Research of War and Peace in Buenos Aires, named after General Professor Golovin. Until his death in 1974, Messner continued to publish works on political and security matters, as well as modern military history, in different Russian émigré publications in Argentina, such as *Russkoye Slovo* (Russian word), *Chasovoy* (Sentinel) and *Nashi Vesti* (Our news). While most of his publications outlined his interpretations of ongoing political and military developments in the context of the Cold War, three of his most significant books—*Lik Soveremennoy Voyni* (The face of contemporary war), *Myatezh—Imya Tret'yey Vseminoy* (Subversion—the name of the third world war) and *Vseminaya Myatezhevoyna* (The worldwide subversion-war)—focused on the conceptualisation of war based on his personal experience and his interpretation of the struggle between the West and Communism (the Soviet Union and, later, China).[17]

Reading Evgeny Messner: avoiding ideological obstacles

When reading Messner's works, it is important to recognise the strong ideological views that informed his writing. Messner held very conservative views, and he was critical of the social and political developments of the first part of the twentieth century. 'Let's see what has been happening in the 20th century':

> The destruction of humanism that tried to replace the spiritual discipline of the Middle Ages with the freedom of the human spirit—it failed; The destruction of socialism (including communism) that was trying ... to discipline the spirit of the human collective and human's soul—this is also failing; The destruction of democracy that was born in the French Revolution and was represented by the then created bourgeois system—which has become senile; The destruction of a culture that peaked in the 19th century—[as today] this culture has been failing to withstand the pressure of decreasing bourgeois perceptions and desires; The destruction of Europe's dominance in the world—the European culture, governance, [and] mores failed to shift the rigid millennial cultures, governances, [and] mores of Asia, to subjugate (not in a colonial way, but in a spiritual one) Africa; The destruction of Christianity that moved away from the Divine Spirit of Christ and adopted formal churchliness, neo-pharisaism and neo-talmudism ...[18]

As a result of his conservatism, Messner often drew unfavourable comparisons between the social norms of the second part of the twentieth century, which he considered an expression of nihilism, and the morals and norms of the past:

> It began in 1913, when the skirts were slightly shortened and a part of women's skin was made visible. These days, even skirts are no longer mandatory, and a woman might enter a church in pants and walk in shorts on the beaches ... In the beginning of the century it was disgraceful to be lustful or behave lustfully, but today it is shameful not to have sex-appeal or not to be receptive to sex-appeal. If it is fashionable, it is not disgraceful—this is the modern law of morals.[19]

In addition to being a conservative, Messner was also a fervent anti-Communist. He believed that 'Red-Moscow' (*Krasnomoskva*) and 'Red-Peking' (*Krasnopekin*) were conspiring together with the aim of dismantling the socio-cultural and moral fabric of Western society, which was too weak to fight back. Accusing the West of Fabianism, pacifism, leftism, radical humanism and liberalism,[20] Messner claimed that these ideas 'are covering Europe, [and] both Americas, conquering the more cultural tiers of less cultural nations':

Lenin, Stalin, Khrushchev broke the culture of Russia. Mao did the same [in China] in an accelerated manner ('The Cultural Revolution'). Mao and Brezhnev are cultivating an absence of culture in the whole world: by using very covert agents or [more] overt ones, like Marcuse ... anarchic-nihilism is implanted, [and] drug addiction and debauchery (all-ages sexuality, homosexuality, pornography) are encouraged ...[21]

Interestingly, in their interpretations of Messner's anti-Communist ideology and conspiratorial ideas, some contemporary Russian scholars have argued that: 'In the end, the Free World, as if it were listening to the theory and recommendations of Messner and other analysts, understood the danger of the Communist Subversion-War and started to "fight back", ultimately achieving victory.'[22] Similarly, one Russian scholar writes that:

> Messner's ideological prejudice did not allow him to construct an objective assessment of a series of events related to the political sources of the subversion-war. Demonising the U.S.S.R. and China, which he considered as the main sources of subversion, he obviously underestimated the role of the U.S.A. and NATO in waging a subversion-war. This [Messner's] conviction might be called his principal mistake, the essence of which has become obvious in our time that is marked by the dissolution of the U.S.S.R. as a result of the Cold War organised by the U.S.A.[23]

In other words, as Evgeny Morozov, one of the contemporary analysts of Messner's works, put it: 'the Messner formula got it right, but in the exact opposite way'.[24] Given that Messner was a product of the times in which he lived, it is thus important to read his works in such a way as to focus on the conceptual characteristics of his theory rather than focusing on his ultra-conservative views. After all, in the words of one contemporary Russian scholar, Messner was a part of:

> The Russian officer corps that by fate's tragic scenario found itself in a foreign land, in an unusual, [and] burdensome situation of refugee and exile ... [These] emigrants created their valuable military-societal works arbitrarily, without any official support ... [and, as a result, these works are characterised by] an unprecedented freedom of military thought (more accurately, the freedom of thought's expression), its uncensored nature, suppleness, full theoretical and ideological pluralism ... [which] engraved their works with maximum criticism and revelation ...[25]

In other words, Messner's works need to be read and understood in their proper context, namely as the writings of a Russian émigré who continued to fight for his motherland, albeit with a pen rather than a rifle, after the 1917 October Revolution, which had stolen so much from him.[26] Although much of his bitterness was directed towards the Bolshevik government, the West was also culpable in his eyes for failing to put up a strong enough fight against Communist ideas.

'The world revolution'

When analysing human history in general, and the history of war in particular, many contemporary Western historians and strategists tend to divide history into several periods characterised by rapid and interconnected socio-political changes. For example, Alvin and Heidi Toffler developed their three-wave theory—agriculture, industry, knowledge;[27] Max Boot suggests four revolutions: Gunpowder, First Industrial, Second Industrial, and Information;[28] and Williamson Murray and MacGregor Knox define five main military revolutions, the second and third of which overlap with each other: (1) the seventeenth-century creation of the modern state and modern military institutions; (2) the French Revolution; (3) the Industrial Revolution; (4) the First World War; and (5) nuclear weapons and ballistic missile delivery systems.[29] Despite the variations in nomenclature, all of the historians just mentioned have essentially observed the same historical pattern, described by Murray as 'earthquakes' in human evolution that bring 'systemic changes in the political, social, and cultural arenas ... [and] recast the nature of society and the state, as well as of military organisations'.[30]

Messner was writing several decades before these works were published, yet he also observed the same historical pattern, which he labelled as 'World Revolution' (*Vsemirnaya Revolutsiya*). In his analysis of the history of the first part of the twentieth century, Messner argues that a chain of socio-political-military changes had brought forth a world revolution that was born 'in both World Wars, as well as in many local ones' and was taking place in six different dimensions: consciousness, morality, social relations, economics, politics and diplomacy (international relations).[31] Similar to Murray, who claims that such socio-cultural–political-military 'earthquakes' do not occur over night and can be traced to lesser, 'pre-shock' changes,[32] Messner argues that:

> The forerunners of the 20th century World Revolution can be found in the 19th century, for example, the Communist Manifesto, the invention of the internal-combustion engine, the rejection-of-violence (*neprotivlenchestvo*) of Lev Tolstoy, the discoveries of Edison, etc. Sometimes ideological, sometimes material, factors that appeared in the last few years of the previous century, at the turn of the century had already become revolutionary and powerful, [thus] creating the World Revolution in this century.[33]

In his analysis of the socio-cultural, political and geopolitical developments via the prism of the six dimensions of the world revolution, Messner accurately points to a range of phenomena, including rising secularism and liberalism; the sexual revolution and human rights movements; the mechanisation of production that released large numbers of people from hard labour, leading to increased social mobility; the rising power of unions defending the social rights of their workers; the rise of the middle class; urbanisation and decolonisation processes; the appearance of powerful international institutions and organisations; and an increase in popular participation in the political process.[34] Thus, what psychologist Steven Pinker would later refer to as the six positive historical trends of human progress (pacification and civilising processes, commerce, feminisation, cosmopolitanism and the escalator of reason),[35] Messner called the world revolution—a period of 'overall madness' in which 'cataclysm took over simultaneously all material and spiritual areas of human life, in all the tribes and nations of humankind'.[36]

Despite his anti-Communist, conservative views, Messner did not hold Communism responsible for creating and spreading the world revolution. Instead, he recognised that such a 'revolution' was the result of the material and social progress that accelerated after both world wars:

> The Communist world thinks that it is required to create the World Revolution. They are wrong: it has already been done. The democratic world makes efforts to prevent the World Revolution. They are [also] wrong: it has occurred already. The division of humanity into two worlds has become misleading: today there is [only] one revolutionised world, in which the Iron and Bamboo curtains separate countries with a more profound Revolution from countries with a less profound [revolution]. And what is regarded as the struggle for-Revolution and against-Revolution is actually a struggle for the deepening or against the deepening of [an already] occurred Revolution.[37]

In Messner's view, the Soviet Union was simply exploiting the social-political and cultural changes created by the world revolution 'like in Japanese jujitsu, where a wrestler uses the strength and weight of his opponent against him, the Red Imperialism uses inimical powers to disrupt the world order'.[38] This idea was echoed in 1989 by William Lind, one of the first proponents of 4GW, who said that, in contemporary war, terrorism employs the tactics of 'judo' by seeking to 'use the enemy's strength against him':

> Terrorists use a free society's freedom and openness, its greatest strengths, against it. They can move freely within our society while actively working to subvert it. They use our democratic rights not only to penetrate but also to defend them-

selves ... Terrorists can effectively wage their form of warfare while being protected by the society they are attacking.[39]

According to Messner, while the world revolution was an undeniable historical fact, the process itself could be slowed down or accelerated as 'life [is] a struggle of two beginnings: progressive and conservative, modification and species' preservation, evolution and conservation ... Evolution is restrained by conservation, conservation is revived by evolution—this is life.'[40] However, regardless of whether fundamental socio-cultural transformations are viewed in negative (Messner) or positive terms (Pinker), the main conclusion to be drawn is that political actors who are able to exploit these transformations for their own benefit will ultimately have the upper hand.

The 'nationalisation of war'

During the 1990s, the proponents of 4GW argued that the character of war was changing.[41] Analysing the social, cultural, political, technological and military transformations since the 1648 Peace of Westphalia, they argued that 'each generation reached deeper into the enemy's territory in an effort to defeat him' and that, in contemporary war, adversaries focus on the 'direct destruction of the enemy's political will to fight' by using society's networks to directly attack 'the minds of enemy's decision-makers'.[42]

Messner had arrived at much the same conclusion more than half a century before Lind, Hammes and other advocates of 4GW. As early as 1931, Messner wrote: 'in our days, humanity experiences a period of social cataclysm' (i.e. the world revolution):

> Every fighting party has the widest opportunity to recruit various allies within the enemy camp. Consequently, a war will be comprised of not only the traditional elements of open war, but also the elements of civil war: sabotage, strikes, unrests, [and] insurgencies will shake the state's organism, proper functionality that is much required at moments of lethal struggle with an external enemy. The enmity will enter the hearts of people, not only the enmity to the foreigners, against whom the war is waged, but also to their kinsmen ... Disputes will undermine the power of the nation and poison [its] soul, making the severe duty of war even more difficult.[43]

In other words, even before the Second World War, Messner had been arguing that attempts to weaken an enemy's political will by influencing the minds of its population would become an integral part of future warfare. It would appear that Messner based this conclusion on his reading of contemporary Soviet strategists, as in a 1926 article published in the German military journal *Wissen und Wehr*:

> The communist military [leaders] start to praise themselves, claiming they have created something new in the military sphere and have established the doctrine of revolutionary war. As Podwoisky writes, this does not consist of making the artillery the main weapon, as Germany did, but [rather] the [power of the] 'word' ... Stepanov proclaims: 'The products of imperialist battles have been mountains of corpses, millions of handicapped and cripples. The products of revolutionary acts of war, by contrast, are to be: annihilation of the inimical life force through propaganda, by means of which they will be turned from enemies into allies. Belitzky recommends identifying the enemy's Achilles heel of social relations, and to work on enemy forces with ideas, mottos, promises, until they are destabilised, but then to administer a short, strong stab against the politically weakest part of the enemy front, which will bring unconditional success.[44]

The 'power of the word' and the role of propaganda in Messner's writing are discussed separately below; here, it is crucial to focus on why Messner argued that this battle for the hearts and minds of the enemy population had become so significant for waging wars. According to Messner, this change in the nature of warfare had been an outcome of the 'nationalisation of war'. Importantly, the word 'nationalisation' is not intended to suggest that war has been shaped by the national characteristics of the involved states; instead, it is used to emphasise Messner's idea that, in contemporary wars, every single member of a given population (i.e. the nation) would be actively involved in the war effort. In other words, according to Messner, waging wars now involves the entire nation:

> Today we have to reckon with the fact that there is no more division between the theatre of war and the country at war, the sum total of the enemies' territory—this is the theatre of war. Today there is no division between the army and the population—all are participating in war with different and gradual intensity and persistence: some fight openly, others secretly, some fight continuously, others only at a convenient opportunity.[45]

Therefore, if, in previous wars, 'a military was breaking an enemy military', and in the Second World War 'a military was breaking an enemy military and its people', in future wars:

> a military and its people are going to break an enemy military and its people: people will be active participants of war, and might be even more active than the military. In previous wars the most important part was considered the conquest of territory. From now, it will be the conquest of the souls in the enemy state.[46]

According to Messner, one of the main outcomes of the world revolution is that people, as nations, have become politically active; or, in other words, the politics of states have become 'nationalised':

> In the beginning of the century each citizen stood alone in face of a powerful government—the [political] parties used to have limited significance ... [Today, however,] a person does not want to be governed, he wants to be self-governed. He does not trust envoys and representatives. ... A nation does not trust its ministers and presidents—it wants to be self-governed.[47]

Thus, due to the socio-cultural and political changes of the twentieth century, people were becoming increasingly involved in the political process. The same also applied in the case of war, with Messner arguing that 'people have stopped being passive observers and silent victims of military struggle':

> A citizen of a free state has got used to a vast opposition to the government ... This predisposes him to oppose the occupying power together with his own military or [equally] to rise against the authority of his country in a union with another fighting party.[48]

In other words, the nationalisation of war (i.e. the increasing role of popular participation in political and military affairs) according to Messner was one of the main outcomes of the world revolution of the twentieth century, and it is one of the main processes that should be exploited to achieve political aims in this new socio-political environment.

The psychological dimension of war and propaganda

Another similarity between Messner's work and the ideas put forward by 4GW theorists three decades later is the conclusion that the transformation of war 'has been marked by greater dispersion on the battlefield'.[49] Whereas Lind, Hammes and other advocates of 4GW divide the history of war into four main generations, Messner uses much simpler language to arrive at the same conclusion. Basing his concept of contemporary warfare on the idea of the 'dimensions of war', Messner argues that 'since the beginning of time, war had been occurring in the bi-dimensional space on land, and, sometimes, on the similarly bi-dimensional sea surface', but 'in this century, however, war has got a third measurement, on land—altitude, and in the sea—altitude and depth'. As a result of this third dimension, 'the deprivations of war, suffering and danger have become the destiny of all people'. If, in the past, a 'population felt itself secure behind its military', today 'there is no reason to trust land or air forces [to secure the population]: they are powerless to prevent the strikes of enemy air forces on the population on the home-front':[50]

> Our wars of 1854, 1877, 1904 and partly 1914 were duels between armies—the population was in the position of seconds. Today, however, the duel has been replaced by a total slaughtering: with the help of aviation, war has transformed from being superficial into [a war with] volume, it got a third dimension, [it] has become a massacre.[51]

On the one hand, the third dimension of war had transformed war into a 'total war', together with 'total destruction, total killing, [and] total madness', in which 'the definition of "military target" has been widened' to such an extent that 'a button factory and a milk farm have become "military industry"'.[52] On the other hand, Messner argued that this total involvement of the population had led to an even more radical transformation of war, as it introduced a fourth dimension—psychological. As war had begun to include the whole of society: 'The soul of the enemy's society has become the most important strategic objective ... Degrading the spirit of the enemy and saving your own spirit from degradation—this is the meaning of struggle in the fourth dimension, which has become more important than the three other dimensions.'[53]

In developing his concept of the psychological dimension of war, Messner argued:

> In the last war [the Second World War], the front line that divided enemies was blurred in places where partisans were erasing it in one or another sides' rear. The future war will be waged not on the line, but on the whole surface of the territories of both adversaries, because behind the military front will appear political, social, economic fronts. The fighting will not take place on a bi-dimensional surface, like in previous times, [or] in triple-dimensional space, as it was since the birth of military aviation, but in four-dimensional [space], where the psyche of the fighting nations is the fourth dimension.[54]

The main outcomes of the world revolution were the politicisation of societies and the consequential nationalisation of war, but 'neither people as a whole, nor a part of them'

> are able to independently think, act or express their feelings. People are a passive power: a power of resistance or a power of sympathy. The crowd is an active power—it is a machine that encloses people's energy; but this machine is not independent ... it is driven. Crowds are controlled by public organisms, different types of communities with leaders on the top ... heroes, or sometimes 'heroes' in inverted commas—made and are making history.[55]

In other words, the increasing involvement of the masses in political and military affairs made them a primary target in war in the psychological dimension, rather than within the physical. With the introduction of the psychological dimension, the main aim of war, according to Messner, is not the physical capture of the enemy's territory, but the psychological conquest of an enemy's spirit in order to 'knock him down from his ideological positions, to bring confusion and discomfiture into his soul'; and the main means of doing so is through propaganda and agitation.[56] Moreover, in his analysis of this phenomenon, Messner focuses on its main characteristics: namely 'propaganda by word' or 'propaganda by deed' and 'offensive propaganda' or 'defensive propaganda':

> The war of the 20th century is not a clear military affair: it consists of politics no less than tactics, the space in this war should be conquered by military, as well as by propaganda. Today nations can deny physical conquest and continue spiritual resistance even after the capitulation of the military. By using propaganda, one should pour the elixir of life into his masses and poison into his enemy's [masses], and by using propagandist antidote [one] should save his [masses] from the enemy's poison.[57]

When discussing the role of propaganda, Messner draws a distinction between 'propaganda by word' and 'propaganda by deed'. Whereas the first included radio, official speeches, publications, theatre, movies and exhibitions, propaganda by deed included successful and timely actions, as 'an idea gets its credibility when its owner supports it with military, political, social, diplomatic, [and] economic achievements'.[58] In other words, Messner argued that propaganda is not only what is said, written, published, broadcasted but also what is done, as 'in times of psychological war, neither victory in battle, nor territorial gains, are the goals themselves: their main value is in their psychological effects'.[59] Furthermore, 'propaganda by deed' is not limited to military activities; it also includes successful political, economic and social actions that can be used to influence the popular psyche, just as a 'successful general strike increases the self-confidence of the working class, [and] the stabilisation of the national currency increases the authority of the government'.[60]

In his discussion of the differences between 'offensive' and 'defensive' propaganda and agitation, Messner argues that while offensive propaganda is intended to weaken the enemy, defensive propaganda is designed to enforce the spirit at home; yet 'it should not be defensive, apologetic [or] justifying, instead it should actively galvanise the emotions and thoughts of our soldiers, warriors and non-warriors'.[61] As such, 'the tone of propaganda should be chosen in accordance with the taste [and] psyche of each nation', as 'defensive or offensive propaganda is doomed to fail if it looks like propaganda'.[62] Thus successful propaganda should be multifaceted—'one half-true for one's own masses, another for the enemy's'—and suitable 'for each level of consciousness, for each category of mores, predispositions [and] interests'.[63]

To summarise Messner's conceptualisation of the role of the psychological dimension of war in general, and of propaganda in particular, it is important to emphasise two elements of his thinking. The first is that the secret of successful propaganda 'is not so much in WHAT to present, but very much in HOW to present [Messner's capitalisation]' as the main tool of propaganda is an 'emotional word' and 'neither healthy logic, nor verity are necessary to create emotions'.[64] The second is the increasing role of the psychological dimension, which allows the states involved in war to manipulate the psyche of each other's population. Hence, in future wars: 'A waging war actor will create and support a partisan movement in the territory of another [enemy] actor; where it will ideologically and materially, by propaganda and financially, support oppositionist and defeatist parties; where it will nourish disobedience, sabotage, subversion and terror.'[65]

As is the case with other aspects of Messner's thought, his conclusion regarding the rise of the fourth (psychological) dimension of war preceded similar conclusions by the proponents of 4GW over thirty years later. Thus, whereas Messner had argued that, with the introduction of the fourth psychological dimension, 'it is easier to degrade a state, rather than conquer it by arms',[66] 4GW theorists would later claim that in fourth-generation warfare the primary goal is 'collapsing the enemy internally, rather than physically destroying him', especially via 'manipulation of the media'.[67]

Messner and the Cold War

With the onset of the Cold War, Messner's understanding of political–military international affairs was influenced by the rivalry between the United States and the Soviet Union. Trying to conceptualise this new situation in the context of a clash of ideologies on the one hand, and the possibility of mutually assured destruction on the other, Messner argued that Trotsky's description of the 1918 Treaty of Brest-Litovsk as 'neither war nor peace' could be said to apply globally after 1945. 'At the conference in Potsdam', Messner argued:

> The international situation had started to be crafted by the formula 'neither peace nor war', getting its most characteristic feature: extremely intense diplo-

matic struggle punctuated by the outbreaks of armed unrests and uprisings. It was called the 'cold war'. It could equally have been foolishly called 'hot diplomacy'. In the formula 'neither war nor peace' there were also some quite 'hot' military operations.[68]

Messner began writing about nuclear deterrence in early 1959. In his view:

> The West should reject the strategic, as well as tactical, planning of war in a nuclear style. The purpose of nuclear weapons should be the mutual compulsion to not use nuclear weapons. The fear of retaliation restrains [one] from committing a crime. Even the most honourless commander would not send assassins to his enemies, because he fears that the same method of struggle will be used against him.[69]

'If in old times',:

> heavy machine-guns, capable of targeting the enemy's heart without fail, had been invented to replace duelling pistols, this would not have led to the abolishment of duels: people would have continued to duel with sabres. The United States and the Soviet Union have their 'heavy machine-guns', [in other words] their nuclear weapons, well aimed at the most vulnerable parts of their enemies. However, these 'guns' are not going to abolish war, but will probably lead to warfare under the threat of the use of hydrogen bombs, without actually using them. In chess, this situation is called 'stalemate'. The United States and the Soviet Union will mutually announce a nuclear-strategic 'stalemate'. Keeping a watchful eye on each other, they will fight in thermonuclear tactics and 'psycho-nuclear' strategy, thus, not by splitting hydrogen atoms, but by splitting ... the enemy's population, their spirit and psyche.[70]

In other words, as a direct military confrontation between the United States and the Soviet Union was unthinkable, both sides would fight each other in the psychological, rather than physical, dimension of war. As part of his argument, Messner interpreted the period's 'proxy wars' as being part of a much broader strategy designed to influence the morale of the enemy's population:

> We must stop thinking that war is when somebody is fighting, and peace is when there is no fighting. The U.S.A., Australia, New Zealand, Philippines and Thailand are not in a state of war against North Vietnam, but they are fighting against it. There is an armistice between North and South Koreas; however, they fight each other, due to the initiative of the North, by partisans on the demarcation line and violent students in Seoul. Israelis and Arabs are considered to be in an armistice, but they are in quite an intensive fight against each other ... It is possible (for the U.S.S.R. and the U.S.A.) to negotiate about non-aggression or disarmament and at the same time to wage a war: the U.S.S.R fights against the U.S.A. by supplying weapons, instructors, money, supplies to those who are fighting against America; and by subverting Americans within the United States.[71]

Analysing this strategic 'stalemate' between the two superpowers, Messner argued that:

> In this era of 'nuclear peace', a classical war (mass armies, their manoeuvres and battles) is unthinkable, as everybody predicts that if it begins with machine-guns and cannons in a big war, then, get excited, someone will use the most scientific weapon, gifted by the best scientist, Einstein, to humanity (nuclear weapons).[72]

And given that the open, 'classic' form of war was no longer possible, a new type of war had begun to develop in its place. This new type, or types, of war he titled as 'half-war' and 'aggressive-diplomacy', which refer to different activities that allow political actors to achieve their political aims without declaring a war. In other words, types of international relations that are 'neither war, nor peace'.

The 'half-war' and 'aggressive-diplomacy'

Messner consequently claimed that there was no longer a clear distinction between the 'state of war' and the 'state of peace'. 'In the old days, it was simple':

> Kutuzov was tasked to defeat Turkey and to sign a peace agreement with it at his discretion, as strategy, rather than diplomacy, had the final word. Clausewitz defined an opposite principle: the aims are defined by diplomacy, and when diplomacy has completed its work, then it is strategy's turn to decide. Thus, the primacy of diplomacy was established. But now the Clausewitzian 'The strategy is a continuation of politics [i.e. diplomacy] by other means' is outdated. It is obsolete because the clear distinction between the phase in which diplomats do their work and the phase when strategists do theirs has been erased. The line between peace and war has been erased. There is no more interchange: peace—war—and peace again. Peace is interwoven with war, war with peace, [and] strategy with diplomacy.[73]

One of the main outcomes of this was the merging together of the 'state of war' and peaceful diplomacy in international relations, and the concomitant creation of new types of relations between states, which fall between cases of open war and declared peace: 'today there are four forms of international relations: war, half-war, aggressive-diplomacy and diplomacy'. While Messner defined 'war' and 'diplomacy' in more or less standard ways as 'an open struggle with weapons' and 'political activities, where the gloves stay on, based on traditional methods of persuasions and threats, begging and extortion', the other two forms, which rest in the middle, were more complex.[74]

Messner was arguing that it was no longer possible to draw a definitive line between peaceful and aggressive relations, especially in the post-Second World War period. If, previously, 'states were living in peace or were waging wars', Trotsky's formula of 'neither war nor peace' created a new situation as 'this formula of refusal to make peace has a new meaning: a refusal to wage an open war'. Consequently, Messner claimed that when 'a defined [and] obvious line between peaceful and military international relations' is erased, different mixtures of neither war, nor peace, start to appear.[75]

Messner described the first new form of relations between states as half-war—'a covert participation in war or feud': 'Red-China behind the Viet Minh fought against France with Vietnam; Washington is waging a half-war against Peking by manoeuvring the 7th fleet in the Formosa Strait; Egypt fights against France in an indirect way—as a revenge for the attack on Port Said—it feeds the feud in Algeria.'[76]

Messner termed the second form 'aggressive-diplomacy', which refers to an 'enforced form of diplomacy' when a country 'does not break the (allegedly) peaceful coexistence with an unappeasable state, but [simultaneously] mobilises oppositionists and revolutionists within it by propaganda and bribery'.[77]

In many ways, Messner's definitions of half-war and aggressive-diplomacy offer little that is new to the study of war and international relations. History is replete with examples of states trying to achieve their goals neither by open war nor by direct diplomatic negotiations, but by something in the middle. For example, King Louis VIII of France politically fed the rebellion in the First Barons' War (1215–17) against the English Crown (before it turned into a dynastic war, i.e. from aggressive-diplomacy to war);[78] similarly, for many years France supported the American colonies by secretly providing supplies, arms and volunteers (most notably the Marquis de Lafayette) before officially declaring war on the British Empire in 1778 (i.e. with the signing of the Franco-American Treaty, the previously waged half-war was transformed into a war).[79]

Instead, the main contribution of Messner's categorisation is that the context of the Cold War, combined with the nationalisation of war and the rise of the psychological dimension in war, made these 'neither war, nor peace' types of confrontations the most prevalent way to achieve political goals. By arguing that in the future the belligerent parties will employ different ways to undermine each other's domestic and international political authority, Messner was one of the first military theoreticians to claim that these types were replacing the 'classic' wars of the twentieth century. Thus, what US strategists would title as fourth-generation warfare in the early 1990s, and what two Chinese colonels termed unrestricted warfare later in the same decade,[80] Messner had conceptualised as *myatezhevoyna* (subversion-war) three decades earlier.

The theory of subversion-war

Before analysing Messner's concept of *Myatezhevoyna*, it is important to clarify some issues surrounding how the term is translated. Some English-language publications (mainly written by East European scholars) have translated *myatezhevoyna* as 'mutiny-war'.[81] This translation, however, is incorrect, as 'mutiny' is an open rebellion against authorities, especially by soldiers or sailors against their officers, the Russian equivalent of which is *bunt*. The direct translation of *myatezh* is 'insurgency', and therefore the direct translation of *myatezhevoyna* from Russian to English is 'a war by insurgency' or 'insurgency-war'. However, there are two reasons why these terms are problematic. The first was given by Messner himself, who argued that there is a difference between *myatezhevoyna* and 'guerrilla war' (i.e. 'insurgency-war'), as the former describes a much-wider phenomenon, and 'guerrilla war' is only one of the possible ways to wage *myatezhevoyna*.[82] The second is that Messner uses *myatezhevoyna* to refer to an activity that is intended to erode an adversary's socio-cultural and military cohesion—an action that is closer to the definition of 'subversion' (*podryvnaya deyatel'nost'*) than 'insurgency'. As a result, in what follows, *myatezh* is translated as 'insurgency', but *myatezhevoyna* as 'subversion-war'.

It is also important to note that, in his analysis of the transformation of the nature of war in the twentieth century, Messner paid special attention to insurgency as a method of warfare:

> Since the times of Napoleon, war had been waged following the rules set by Napoleon, Clausewitz, [and] Delbrück (in the Russian version—by Jomini, Leer, [and] Mikhnevich). Since Ludendorff (1916), war had been fought utilising the ideas of the economist Adam Smith, supplementing the confrontation of armies with the struggle of economies. Stalin and Churchill (both neither generals, nor military people) added to Clausewitz and Smith the theses of the civil theoreticians on military art—Marx, Engels, [and] Kautsky—who gave the main roles in waging wars not to the armies or economies, but rather to insurgencies. Since the beginning of the Indochina War ... the participation of armies in military confrontations has become less significant: Clausewitz and Engels were pushed away by the heralds of the new type of war, Giáp, Mao, [and] Che Guevara.[83]

In Messner's view, the military conflicts that took place in a twenty-five-year period after the Second World War were all linked, as 'one case of an infectious disease is an accident, but thousands of simultaneous cases are an epidemic'. This epidemic was comprised by a large number of seemingly unconnected confrontations, each one of which had broken out for different reasons (religious, racial, social, ethnic, tribal, ideological, political, etc.); however, they all shared 'one mutual aim—the destruction of the World, which has become outdated and does not know how to exit the cul-de-sac that it [the world] created by itself'.[84] Owing to his conservatism, it is hardly surprising that Messner attributed these seemingly independent conflicts to what he described as the world revolution. Yet while he was opposed to what had taken place as part of the world revolution, a process in which 'inter-cultural change is galloping',[85] it was important to understand how it had happened and how it could be used to achieve political goals:

> There are many incomprehensible occurrences in the world, [especially] if we look through the prism of the outdated definitions of war; however, a perspective through the new lens of Subversion-War might explain a lot. Then we will stop calling strategic actions that are part of Subversion-War, as criminal occurrences ... We will stop calling operational and tactical episodes of Subversion-War as [independent] unrests.[86]

The nationalisation of war resulted from the world revolution, and, therefore, the involvement of the masses (politically and militarily, directly and indirectly) in war-making, as active participants or as targets, had increased: 'Today, the regular forces have lost their military monopoly; the irregular forces are fighting together with regular (or, even more than regular) [forces].'[87]

While the prevalence of insurgencies over 'classic' wars was an uncontrollable and inevitable process in the context of the world revolution, it could also be exploited by political actors to achieve certain political gains by purposefully encouraging internal conflicts, which would in turn create '"problems" and crises' that will 'gradually estrange [the] masses from [their] government'[88]—or, put differently, undermine an adversary's political legitimacy. This 'subversive' type of warfare, Messner argued, is 'psychological warfare aimed to conquer the mind and soul of people, [of] an attacked nation';[89] thus, one of the most distinctive characteristics of a subversion-war is the prominence of the psychological dimension:

> This new phenomenon has to be considered from different perspectives, but the most important one is psychological: if in classic wars the morale of standing armies was of a great importance, then in the current era of nations in arms and

violent popular movements, psychological factors have become dominant. A people's army is a psychological organism; therefore, a popular movement is a purely psychological phenomenon. A [combined] war of military and popular movements—subversion-war—is a psychological war.[90]

'When war was a tournament—an army against an army':

it was relatively easy: find a large field and fight to destroy the enemy's formation, try to break force by force. Today, however, in times of psychological war, neither victory in battle, nor territorial gains, are the goals themselves: their main value is in their psychological effects. [Therefore] one should think not about the destruction of an enemy's manpower, but about the crushing of his psychological power. This is the surest way to victory in a subversion-war.[91]

While insurgencies around the world might be considered as unrelated, chaotic events, 'this chaos is created not chaotically, but quite systematically, organised, [and] thought-out by the guiding strategic centres',[92] which coordinate warfare in the psychological dimension to accelerate the internal contradictions created by the world revolution with the aim of weakening a political adversary. Describing the complex nature of this coordination, Messner argued that:

Waging war is an art. Waging an insurgency (revolution) is also an art. Today, a new art is developing—waging a subversion-war. Strategists almost always face difficult choices in defining purposes for their actions (interim and final). In a subversion-war the choice is especially difficult, due to the abundance of goals and differences in their significance (either purely psychological, or material with a psychological side, or purely material).[93]

These goals could be placed in a hierarchy:

1) the dissolution of the spirit of the enemy public; 2) the defeat of the enemy's active part (the military, partisan organisations, and violent popular movements); 3) the seizure or destruction of objects of a psychological value; 4) the seizure or destruction of objects of material value; 5) a creation of an impression of order to obtain new allies and crush the spirit of the enemy's allies.[94]

There is little new in this sequence of goals, as Messner acknowledged that subversion-war followed the steps formulated by Mao Zedong: 'demoralisation, riots, terror, gradual recruitment to rebellious [movements], reconstruction of [people's] souls (creation of a "new human"), [and] construction of a [new] human-machine system (social)'.[95] Instead, the main innovation in Messner's hierarchy is how the first step was to be executed.

Due to the world revolution, 'wars had merged with subversions, subversions with wars, creating a new form of an armed conflict, which we will call subversion-war, and in which fighters are not so much the troops themselves, but public movements'.⁹⁶ As the world revolution is a global and uncontrollable phenomenon, different public movements might pursue a similar goal, but their participation will be chaotic and uncoordinated. For example, in his discussion of the Vietnam War, Messner argues that the main problem for the United States was its inability to realise that, in addition to the direct war with Communist Vietnam, it was fighting a range of subversions led by:

> African-Americans headed by Stokely Carmichael; 'doves' in Congress; student anarchists in Europe headed by Dutschke, Teufel [and] Cohn-Bendit; [and] pacifists from a variety of countries, led by the Pope. The World Council of Churches and Lord Russell, 81 Communist parties [around the globe] and the U.S.S.R., which is unreachable to Johnson's 'friendship bridges'. All these powers wage battles attacking the spirit of the White House by a world-wide subversion and weakening the belligerent volition of the U.S.A.⁹⁷

For Messner, victory in future wars would depend on exploiting the socio-cultural changes created by the world revolution.⁹⁸ Thus, at the heart of the idea of subversion-war, is the notion that political actors can exploit these changes by actively supporting domestic confrontations created by the world revolution with the aim of subverting their adversaries and demoralising the spirit of their citizens, thus eroding an adversary's political legitimacy.

It is important to note that Messner did not claim that a subversion-war has to involve direct physical or psychological actions against an enemy. Though similar to 'classic' wars in certain respects, a subversion-war can be fought against an adversary by actively supporting the opposition within the enemy state; the true art of subversion-war, according to Messner, involves indirect and seemingly unrelated actions that encourage internal confrontations. Such opportunities should be identified by 'psychological reconnaissance': 'there are no more delicate and complex tasks, than those of psycho-reconnaissance: [while] a doctor psychoanalyst deals with one deviant soul, psycho-reconnaissance [deals] with the souls of nations'.⁹⁹ As subversion-war is waged first and foremost in the psychological dimension, it is important to understand the pre-existing socio-cultural factors that could undermine an adversary such as the USSR, which, '[while] disseminating subversion, does not provoke a big war, [and] does not incur upon itself a Western crusade'.¹⁰⁰ The best example of such a subversion-war, according to Messner, was the Vietnam War, where one of the main battlefields was between 'Red-Moscow and Washington (on the territory of the U.S.A.), where defeatists, traitors and betrayers, agents of the Red-Kremlin or its sympathisers tried to fight for the victory of North Vietnam.'¹⁰¹ Another apposite example was the Israeli–Arab conflict, where the purpose of Moscow's subversion-war was 'to continue in Palestine the economic and moral attrition [of the United States] that started in Vietnam'.¹⁰²

In summarising Messner's concept of subversion-war, it is important to highlight its two main characteristics. The first is that it does not simply support political disorder, opposition movements or insurgencies, but uses them directly or indirectly to achieve political aims, as was the case with the Soviet Union's support of North Vietnam in order to prolong the war, 'hoping for a victory, not in Vietnam, but in the U.S.A.,' where the actual main battle was being waged within US society.¹⁰³ The second characteristic of subversion-war is its unpredictability. Since it is waged in the psychological dimension, and uses ongoing socio-cultural and political contradictions, 'sometimes it is difficult to predict the path the psychological pseudo-logic [will take]'.¹⁰⁴ In addition, the complexity of the different strategic aspects that are involved in subversion-war in the context of world revolution and the consequent nationalisation of war adds even more unpredictability:

> The traditional military objective—the enemy's army—has been dispersed into people at war; the traditional geographic objective—the capital—has become one of [many] surrogate capitals, i.e. different centres of national infrastructure. Vilnius–Smolensk–Moscow—this was the trajectory of Napoleon's strategic arrow; Moltke's arrow was directed from Sedan to Paris. Today, the strategic idea might be like a ball, shot by football players from one side of the field to another: there is one final goal, but there are also many intermediate ones, and they are in different fields: military, diplomatic, economic, socio-political and physiological.¹⁰⁵

The revival of Messner in post-Soviet Russia

Due to his anti-Communist views and alliance with the White Movement, and later with Nazi Germany, Messner remained generally unknown in the Soviet Union. His books and articles, similar to the works of other Russian émigrés published abroad, probably reached the USSR, but access to them was restricted to a small number of high-level officials and professionals. In the post-Soviet period, however, Messner's works have become available to a broader range of military thinkers, and there has been a growing revival of Messner's concept of subversion-war to analyse the contemporary geopolitical situation and political, military and economic confrontations (e.g. the conflict in the Balkans, the rise of terror organisations, the Arab Spring, the Ukrainian Crisis).¹⁰⁶

There are two reasons for this revival. The first is the general revival of the works of Russian émigré military thinkers immediately after the end of the Cold War. With the collapse of the Soviet Union and the degradation of the Russian military physically (due to the division of the Red Army between fifteen new states) and intellectually (due to the loss of Soviet ideology), Russian military thinkers began looking for a new philosophy of war. As a pair of Russian officers stated in 1994: 'By rejecting the exclusive role of Marxism as the solely true teaching that explains the nature and character of war ... we put ourselves in the face of necessity to clarify the [military] scientific basis of [our] worldview, our views on war as a special societal condition.'[107]

Thus, with the disappearance of Marxism–Leninism as a state ideology in the 1990s, which previously restricted access to the Russian émigré military and political thinkers, and the thirst for a new perspective on the phenomenon of war, members of the Russian military became increasingly interested in this literature. Colonel Alexander Savinkin, the founder and main editor of *Rossiyskiy Voyennyy Sbornik* (Russian military collection)—a series of books supported by the Russian Military University and the Russian Ministry of Defence—edited at least nine books that republished and analysed the works of Russian émigré military and political thinkers.[108] In his words:

> During their life-time, the White émigrés did not and could not solve their political tasks, but they left a great spiritual legacy, values and guidelines for the development of a patriotic education for the future of the armed forces of liberated Russia. [And therefore] The spiritual revival of the military today is impossible without the studying and understanding of the patriotic and military thinking of the Russian émigrés, the ideas of the White [Movement] and [its] traditions.[109]

In addition to this series of books, analysis of the ideas produced by the Russian military émigrés has now extended to include a vast number of books[110] and articles,[111] all which emphasise that:

> The phenomenon of Russian military culture abroad is unique in modern history ... Obviously, many statements and ideas are very different from those we were guided by for many years. But we must remember that this heritage is based on a bitter historical experience, infused with the blood and tears of our suffering Motherland. The works of the best military minds in exile are dedicated to her, hoping to be useful for the 'future' Russian Army.[112]

The second factor behind the revival of Messner's theory is the fact that his ideas directly correspond to the conceptualisation of contemporary Russian military thinkers on the changing character of war in the post-Soviet era: 'A number of Russian [military] experts have considered that in the future, wars will be waged not as classic wars and military operations, but by strategy of indirect actions (subversion-war), where major roles will be played by terrorism, unlawful military formations, informational-psychological means of influence, etc.'[113]

As the starting point of this conceptualisation was the analysis of the defeat of the Soviet Union in the Cold War—'the biggest geopolitical catastrophe of the 21st century', according to Major General Alexander Vladimirov[114]—it is not surprising that this analysis ultimately led some Russian military thinkers to conclude that: 'The example of the Cold War demonstrates that the U.S.S.R. was the first and most significant victim of subversion-war, proving the necessity to broaden the spectrum of war criteria.'[115]

Or, in the words of one Russian colonel,

> The essence of the Cold War was a purposeful, organised, coordinated, uncompromising, and complex interstate and inter-coalition non-military struggle ... It is obvious that only an inability (or unwillingness) to achieve victory ... by implementing the means of armed struggle creates the necessity in other methods, means, leverages and channels, which would be effective enough to replace [armed] violence and force the adversary to accept its defeat without [actually] fighting.[116]

Summarising this analysis of the Cold War and the role of non-military means and methods in contemporary conflicts, Professor Ivan Radikov claims that:

> If in traditional wars the main priority was the physical dimension (military potential, armed forces, the economic and demographic potential of the adversary), then, since the beginning of this new century, the frame of the struggle has drifted towards mental (the consciousness of the elites, the consciousness of the masses and their psychological state, the means of mass communication) and spiritual (the religious consciousness of people, public morals, inter-religious and intra-religious relations and the traditional religious system) dimensions.[117]

While the development of this broader understanding of the war phenomenon was mainly based on analyses of the Cold War,[118] the conceptual interconnection between contemporary military thinking and the theory of subversion-war can easily be identified by the numerous references to Messner's works in the Russian literature.[119] Moreover, Messner's idea that 'in our times it is easier to degrade a state, rather than to subdue it'[120] has found its way into two major theories on contemporary wars that were developed in Russia in the late 1990s and early 2000s: 'net-centric warfare' and 'information war', which lie at the core of the following chapter. Although these two theories were conceptualised independently of each other, as well as from Messner's concept of subversion-war, they share many aspects and assumptions, all of which eventually shaped the Russian conceptualisation of *gibridnaya voyna*.

4

NET-CENTRIC AND INFORMATION WARS

MODERN THEORIES OF SUBVERSION

As mentioned in the previous chapter, in addition to the revived interest in Mesner's work, two other prominent theories of war—net-centric and information wars—were also developed during the same period by two Russian scholars, Aleksandr Dugin and Igor Panarin. Although neither mentions Messner's works, the essence of their concepts echo many of Messner's ideas. Indeed, the core of Dugin's net-centric war and Panarin's information war (as well as Messner's subversion-war, as it was interpreted by many Russian contemporary military thinkers) is quite similar. All these theories claim that in the geopolitical and technological realities of the twenty-first century, it is easier to achieve political goals by undermining the political authority of the adversary by manipulating political elites and generating political dissent, separatism and social problems, rather than by waging classic wars and military operations.

Aleksandr Dugin: the theory of net-centric war

Aleksandr Gelyevich Dugin is a Russian political scientist, geopolitical philosopher, religious historian and a Slavophil. He started publishing his work in the late 1980s and has proved to be a talented writer and speaker. In the period since, he has established himself as a prolific author, publishing several books (almost one every year), as well as hundreds of articles, commentaries and interviews, some of which have been translated to English and other languages. Dugin has also held several senior advisory positions in the Russian political establishment, and from 2009 to 2014, he served as the head of the Department of Sociology of International Relations of the Lomonosov Moscow State University.[1]

A full discussion of Dugin's ideas is beyond the scope of the present chapter. However, it is important to highlight several aspects of his ideas that are relevant to the current discussion. First, Dugin is one of the most prominent advocates of the idea of a Russian Eurasian civilisation, an idea designed to encapsulate Russia's unique socio-cultural characteristics, history and role in the global arena.[2] Thus, Dugin argues that:

> Russian society is Eurasian, that is partly European and partly non-European, and, [therefore] it is generally a unique and distinctive phenomenon ... We live on this land, within these boundaries not by accident. These borders are also inhabited and settled by us not by accident. Between them and us there is a direct sociological, cultural, genetic, causal, conceptual, [and] morphological relationship.[3]

Secondly, Dugin holds a strong, ideologically informed position on relations between Russia and the West, arguing that Russian Eurasian civilisation has consistently been faced with Western aggression, chiefly represented by the United States: 'the U.S.A [is] the sum of the West, its political, religious and ideological vanguard ... [it is] the incarnation of the West, of Western capitalism, its centre and axis, its essence'.[4] According to Dugin, Russia has always been one of the most vital enemies of the West, and this struggle, between Western Protestant civilisation (led initially by the British Empire and then by the United States) and Russian Orthodox Eurasian civilisation, can be traced throughout hundreds of years of confrontations:

> from the mid-20th century, the geopolitical duel, which has been traced by geo-politicians [all the way] down to the ancient conflicts of Athens and Sparta, Rome and Carthage, etc., finally crystallised into the collision between the Western world (the U.S.A. and Western Europe) and the U.S.S.R. with its satellites in Europe and Asia.[5]

The third and most relevant aspect of Dugin's work concerns how, in his opinion, the West (mainly the United States) has been waging its aggression against Russia in the twentieth and early twenty-first centuries. In 2007, Dugin published a book entitled *Geopolitika Postmoderna* (Geopolitics of postmodernism), in which he presents his interpretation of the US concept of 'network-centric warfare'.[6] Following this book, Dugin continued to clarify his ideas in a series of articles[7] and other books,[8] introducing the term net-centric war to a Russian audience. As Dugin's interpretation of this concept differs substantively from its US counterpart, as well as how the original US term was discussed within the Russian military, it is important that these differences are clarified before proceeding with the discussion.

The concept of net-centric warfare is a product of US military thought. It first appeared in the late 1990s in the publications of the US Navy.[9] In 1998, reflecting on the social, economic and technological changes that had been occurring in the last decade of the twentieth century, Vice Admiral (ret.) Arthur K. Cebrowski argued that: 'Arising from fundamental changes in American society and business, military operations increasingly will capitalise on the advances and advantages of information technology.'[10]

The concept of net-centric warfare was further developed and promoted by a group of experts from the Command and Control Research Program (CCRP) within the Office of the Assistant Secretary of Defence, who derived their theory from a series of case studies on how businesses were using new information and communication technologies to improve situation analysis, projecting the very same concepts on to military activity.[11] According to them:

> NCW [net-centric warfare] is about human and organizational behaviour. NCW is based on adopting a new way of thinking—network-centric thinking—and applying it to military operations. NCW focuses on the combat power that can be generated from the effective linking or networking of the warfighting enterprise. It is characterized by the ability of geographically dispersed forces (consisting of entities) to create a high level of shared battlespace awareness that can be exploited via self-synchronisation and other network-centric operations to achieve commanders' intent.[12]

In other words, the main idea of net-centric warfare, according to this original conceptualisation, is to enhance the effectiveness of military units on the battlefield by increasing the efficiency of the collection, aggregation, analysis and communication of valuable and relevant data from a large number of sensors: 'Empowered by knowledge, derived from a shared awareness of the battlespace and a shared understanding of commanders' intent, our forces will be able to self-synchronise, operate with a small footprint, and be more effective when operating autonomously.'[13]

Whereas Russian military experts had originally conceptualised net-centric war in a similar way to their US counterparts,[14] in Dugin's work the same concept is conceptualised in a much broader way. Thus while US and Russian military thinkers simply claimed that adapting net-centric warfare concepts to military operations would enhance the effectiveness of deployed forces, Dugin claimed that net-centric warfare resulted from a much broader transformation, as a 'network [itself] is a fundamental and absolute phenomenon, a development of which alters the political, economic, social, cultural and anthropological picture of the world',[15] and, therefore, an analysis of net-centric war should not be restricted just to the military realm.

Dugin does not restrict his interpretation of net-centric war to military terms, as his definition of network extends well beyond the sphere relevant to military operations per se. '[A] Network', he writes, is an 'informational dimension, in which major strategic operations are developed, [including] their media, diplomatic, economic and technical support', and it includes different elements that were previously considered separate, such as: 'Military units, communication systems, operational information support, the management of public opinion, diplomatic measures, social processes, intelligence and counter-intelligence, ethnopsychology, religious and collective psychology, economic support, academia, technological innovation, etc.'[16]

Basing his conceptualisation on a general division of human civilisation into three periods—agrarian, industrial and informational (pre-modern, modern and postmodern)—Dugin claims that a 'network' is a fundamental characteristic of the contemporary informational phase of the civilisation process, and 'the theory of net-centric wars is [just] a projection of major aspects of postmodern development in the sphere of military science'.[17] Therefore, according to Dugin, while the aims of net-centric war have remained similar to traditional wars, 'establishing control over an area, which does not belong to the fighting actor, or keeping control over an area, which does belong [to this actor], from the adversary's encroaching', the new 'postmodern' realities have created an entirely new understanding of what an 'area' is, what 'control' means and who the 'adversary' is:

> In net-centric war, reality is secondary in relation to virtual. The image is much more important than reality. Reality itself becomes real only after reports about it appear in the informational dimension, and therefore, the major factor is control of the informational dimension. The one, who controls the informational dimension, controls everything. Informational support of war stops being a secondary supporting factor (as classic propaganda was), but becomes the raison d'etre of war. In essence, a war has become informational, [where] military operations have only a secondary supporting role.[18]

The first characteristic that distinguishes 'postmodern' net-centric wars from traditional (i.e. 'modern') wars is the 'area' over which war is waged. According to Dugin, this area is not a physical territory with strictly defined borders, but a virtual dimension created and represented by a network of interconnected informational trends, as 'in net-centric wars, an occupation or annexation of territory are not required, [because] an establishment of control over a network will be enough, as it intends to control mass media, financial instruments, access to technologies, political and cultural elite, centres of active youths'.[19] In other words, as a 'network' is a social phenomenon represented in the informational dimension, net-centric warfare is when a political actor tries (not necessarily by military means) to control a stronger and wider network, undermining the network of his adversary and simultaneously protecting its own from the adversary's encroachment.

The second distinctive characteristic of net-centric war, according to Dugin, is the way in which this 'network' can be controlled, or, in other words, how such a war can be fought. Since the main defining factor of a network is information, and as the amount of accessible information is constantly increasing, the main goal of the parties involved in a net-centric war is not to control the information itself but to rule and manipulate its nature. By controlling how and which information is created, aggregated and shared, 'the experts of network strategies can give even to very negative and dangerous information ... an entirely opposite character, making it harmless or fading its impact'.[20] As the essence of control in net-centric war is not about the physical dimension but about the informational (virtual) one, the main role is given to how information is deciphered, interpreted, structured and presented to enhance one's control over one's own network, simultaneously undermining the control of one's adversary.

This leads to the third characteristic that distinguishes net-centric warfare from its US counterpart, namely the nature and identity of the adversary. As the nature of a network is very dynamic and flexible, and 'today's ally can become tomorrow's adversary (and vice versa), the separation between friends and foes, in this new type of wars, is very provisional'.[21] Therefore, net-centric operations are waged consistently against everybody, 'shaping the behavioural model of friends, neutral parties and enemies in situations of peace, crisis and war'.[22] This ultimately leads to a more dynamic image of an adversary, which is defined more in virtual terms than physical ones. Moreover, since net-centric war is a war in an informational (virtual) dimension, it might even be waged against a virtual (fictional) adversary, even if the consequences of such a war are very physical:

> Al-Qaida and international terrorism present an unlocalisable and untraceable power, under the definition of which at any time it is possible to describe entirely

different realities: once it is the Taliban in Afghanistan (which is related to a certain degree to the Salafist movements of 'pure Islam' like Ben-Laden's organisation), another time it is Hussein's Iraq (which had neither ideological, nor organisational relations with Al-Qaida), or Shiite Iran (which is a direct political, religious, ideological and organisational opposite to Al-Qaida).[23]

In other words, this type of confrontation focuses on control over the aggregation, interpretation, deciphering, manipulation and distribution of information to relevant active groups (i.e. the network) in an attempt to achieve certain political aims. Dugin points to different types of networks. The first type is a 'natural network' made up of ethnic, religious (and to some degree sexual) minorities. These groups of minorities usually lack an official political status, existing de facto and not de jure, which 'allows them to act without attracting [the] direct attention of law and legal procedures, [as] from the legal perspective the representatives of these groups are common citizens, just like anybody else, and usually they require no registration to maintain and develop their network's connections'.[24] Independently, these groups are generally unable to achieve strategically important goals, yet in net-centric warfare they are the first and major target of foreign political actors, who try to establish their network and undermine the regime's existing network by influencing and manipulating these groups: 'by investing attention, resources and technical skills in key players of these networks, it is possible to achieve colossal successes'.[25]

The second way to establish a network able to facilitate the desired flow of information is by creating 'artificial networks' that would complement the natural ones. By artificial networks, Dugin is referring to different non-governmental organisations (NGOs) such as 'human rights movements, non-commercial partnerships, educational initiatives, centres that support research activities, academic and social networks, public organisations, etc.'[26] As these NGOs generally behave in law-abiding ways, it is difficult to recognise when they are acting as foreign agents, facilitating and promoting information based on the values of the foreign network and undermining the existing one.

These two types of networks (natural and artificial) lead to the third way of shaping a desired flow of information—the creation of 'agents of influence'. It is important to note that, in Dugin's interpretation, this term loses its traditional meaning (i.e. recruited or indoctrinated agents) and refers to a much broader phenomenon, namely any political actor that produces or facilitates information that can be used to enhance the network. These agents of influence obviously include those who intentionally (due to ideological beliefs or other self-interests) or sincerely (without knowing that the organisation they are part of constitutes a foreign network) promote the values and interests of the foreign network. Yet these agents of influence might include not only those who promote and defend the values of the foreign network but also those who openly reject them, as 'similar to yachting, one might move forward despite the front wind, influencing only small actions and piecing the rest together from an accurate and synchronised presentation of information'.[27] In reference to the previously cited example of Al-Qaida, Bin Laden ultimately became an important agent of influence despite opposing Western values, as his image and actions, skilfully shaped in the information dimension, were used to enhance the Western network and not to undermine it.

The fourth and final way of establishing and influencing networks to facilitate and direct the desired flow of information is the globalisation process, as 'the very fact of the inclusion of economic, energetic, informational [and] academic resources of a state into global networks automatically offers advantage to those who control the code, the protocol and the algorithm of the functioning process of these networks'.[28] While in the informational dimension globalisation is presented as an 'objective', 'positive', 'progressive' process that leads to 'development' and 'modernisation', in reality, Dugin claims, it 'increases the risks of the establishment of external control, as the architects, who created and control these global networks ... are in a much more advantageous situation, rather than those who just connect to them'.[29] Going even further, Dugin suggests that: 'Globalisation and its most vivid institutions: WTO, IMF, the World Bank, etc., are one of the forms in which strong participants of the global process (first and foremost the U.S.) wage a net-centric war against the weaker ones.'[30]

This leads to the ideological component of Dugin's conceptualisation of net-centric war. As this new type of war was, according to Dugin, developed and honed by the United States, he argues that the main purpose of US net-centric warfare is to establish 'full and absolute control over all participants of actual and possible military activities, and their [participants'] total manipulation in all situations—when a war is waged, when it only matures or when there is peace'.[31]

This might be achieved, Dugin claims, by winning an absolute superiority in the informational dimension because, as discussed above, it has a 'highly important, if not central, role' in net-centric wars, creating 'the most prevalent environment of network wars, which has evolved into an independent domain—"info-sphere"—that stands separate and equal to the physical dimensions'.[32] For example, one of the main purposes of US net-centric war, according to Dugin, is to inculcate in everyone's minds that military competition with the United States is pointless and should be avoided at all costs. Through the informational dimension, Dugin claims that Washington has been seeking to:

> build a system of the global domination of the U.S.A. over the whole world, i.e., the postmodern analogy of colonialism and submission, executed under new conditions, in new forms and by new means. There is no need for direct occupation, a massive deployment of the forces or territorial conquest ... 'Network' is a much more flexible weapon, it manipulates with violence and military power only in extreme cases, [while] the major results are achieved by contextual influence with a wide aggregation of factors: informational, social, cognitive, etc.[33]

As the transition to the postmodern (i.e. post-industrial, informational) age has been led by the United States, Dugin argues that, beginning in the late 1970s, the US approach to conflict started to evolve towards networks, and the Soviet Union was the first target of this new way of war. 'The U.S. could not beat the U.S.S.R.', he claims:

> neither in a direct confrontation, nor in a direct ideological battle, nor in any direct way of a struggle between special services ... Then the major principle of networking strategies was employed: informal infiltration [focused on] finding weak, indeterminate, entropic elements within Soviet hierarchy. The U.S.S.R. was defeated neither by a counter-power, nor by an anti-Soviet organisation, but by skilfully organised, manipulated and mobilised 'entropy'.[34]

As the fall of the Soviet Union was an outcome of the Soviet defeat in the net-centric war waged by the West (mainly the United States), the Western network's penetration of Russian society reached a peak in the 1990s. During that period, the West actively incorporated different Russian social, economic, academic and other groups of influence into its network. Dugin lists these as follows:

Financial system;

Economy (through oligarchs);

Political elite (through ideas of liberalism and pro-Westernism);

Mass Media (through copying Western shows);

Educational standards (through the imitation of Western education systems and adaptation of tutorials written in the West or according to Western values);

Scientific Institutions (through a system of brain exportation or provision of Western grants);

Youth environments (through fashion, images and the Internet).[35]

The synchronisation of these groups with the Western (US) network, Dugin continues, ultimately led to the economy and politics of Russia being placed under foreign control in the late 1990s, as well as 'the passiveness and intellectual inadequateness of the highest political administration, [which] put the state on the edge of dissolution'.[36]

Dugin claims that this situation changed with Vladimir Putin's ascension to power, as he immediately started acting to restore the sovereignty of Russia's traditional networks, thus undermining the influence of the Western network in Russia. During his first eight years in office, Dugin writes that:

> Putin gradually isolated and neutralised the most dangerous segments of the Atlantic networks: oligarchs, pro-Western liberals, the promoters of American interests, NGOs, which were directly involved in spying activities. Putin broke down the [subversive] processes within ethnic minorities by making cooperation with the government more attractive, [thus] bringing Chechnya back into the folds of Russia.[37]

Yet, despite his positive appraisal of Putin's actions, Dugin concludes that they were not an outcome of an understanding of the concept of network and the theory of net-centric warfare, but rather a result of Putin's immunity to the influence of the Western network; thus his actions were guided more by intuition, rather than a consciously developed plan. Therefore, despite Putin's success in overcoming the most critical phase of the net-centric war waged by the West against Russia in the first decade of the twenty-first century:

> None of the [Western] network structures, which have been acting on Russian territory, were completely eliminated. Moreover, new technologies have been developed and tried all the time. Some skills in waging informational campaigns [i.e. net-centric operations] we [in Russia] have learnt from the adversary and now are enjoying them. But nothing stands still. And if we will slow down our net-centric war on behalf of Eurasia, the critical situation might precipitously come back.[38]

Thus, in sum, the purpose of net-centric warfare is to influence networks of people, instructions, foundations, organisations and so on that intuitively (or otherwise) promote a certain set of ideas in an attempt to achieve certain political goals. The United States was the first to master this new type of warfare, waging its net-centric wars against all other countries and nations by manipulating their social processes from the inside, thus winning physical confrontations before they had even begun. Therefore, Dugin claims, if Russia does not 'postmodernise' its military, secret services, political institutions, information and communication systems to suit this net-centric informational struggle, it is doomed to lose this war.[39] According to Dugin, the United States has been waging a persistent and carefully planned offensive on Russia in the informational domain as part of the US strategy to dominate the world in the postmodern informational age. This net-centric war is being waged against Russia by a carefully crafted network that includes:

> a Pro-American lobby of experts, political scientists, analysts, [and political] technicians that closely surround [Russian] authorities. A vast number of American foundations [that] actively operate connecting the intellectual elites to their network. The representatives of Russian capital and senior officialdom [that] are naturally integrated within the Western world, where their savings are kept. The means of mass communication [that] irradiate readers and viewers with cascades of visual and semantic information, built according to the Western patterns.[40]

To withstand this attack, Dugin argues that Russia has to adopt the 'Eurasian model' by taking a position in opposition to the 'Atlantic-American model', thereby creating its own network, which will be oriented in precisely the opposite direction: 'We will be able to defend Russia, to enforce and preserve Eurasia as a pillar of the multi-polar world, only by being consciously prepared to engage in the net-centric war (which is waged against us anyway) and [thus] winning it.'[41]

This Eurasian network, which would offer a symmetric response within the informational dimension, could be based on: 'Special groups that would include senior officials, the most passionate cadre of different special service agencies, intellectuals, scientists, engineers, political scientists, [and] corps of patriotically oriented journalists and culture activists.'[42]

While Dugin's conceptualisation of net-centric war has not been embraced by the Russian military community, his ideas have been adopted by a large group of Russian political scientists attracted to his interpretation of the conflict between East and West (in this case, Russia versus the United States). Since being introduced in 2007, Dugin's concept of net-centric warfare has been used to interpret the different geopolitical events in the post-Soviet arena and worldwide in such a way as to suggest that they are all parts of the same net-centric war waged by the United States for global domination, and first and foremost the defeat of Russia.[43]

Igor Panarin: the theory of 'information warfare'

Igor Nikolaevich Panarin holds a higher doctoral degree in political science and a PhD in psychology; he is a full member of the Military Academy of Science of the Russian Federation, as well as holding numerous senior advisory and coordinating positions in the Russian political system. Since the mid-1990s, Panarin has published more than twenty books, hundreds of articles, commentaries and interviews, most of which focus on the psychological facets of warfare in general, and information warfare in particular. Whereas Dugin focuses primarily on the struggle between the West and Russia from the perspective of a net-centric war that occurs within the informational dimension, Panarin instead focuses on what he calls 'information warfare' as the major domain of this struggle:

> Since antiquity, the stability of the political system of any country has relied on how quickly and completely the political elites receive information (e.g., about [possible] danger), and how quickly they respond ... Political activity [by definition] is an informational struggle over the control of the minds of the elites and [other] social groups.[44]

In his analysis of the history of warfare, Panarin claims that the informational dimension has consistently played a decisive role in human conflict.[45] In Panarin's words, information war is:

> A type of confrontation between parties, represented by the use of special (political, economic, diplomatic, military and other) methods [based on different] ways and means that influence the informational environment of the opposing party [while] protecting their own [environment], in order to achieve clearly defined goals. [Therefore] The major dimensions for waging informational-psychological confrontations [are] political, diplomatic, financial-economic, [and] military.[46]

It is important to note that, when Panarin mentions these dimensions, he does not refer to direct political, diplomatic, financial-economic or military activities, but rather to the manipulation of their informational images in order to control the targeted public opinion in order to gain certain political benefits. This control can be achieved by information manipulation, disinformation, fabrication of information, lobbying, blackmail or any other possible way of extracting the desired information, or simply by the mere denial of information originating from the adversary. Thus, when an information war is waged by one state against another, Panarin states that it 'aims to interrupt the balance of power and achieve superiority in the global informational dimension' by targeting 'the decision-making processes of the adversary' via the manipulation of international and domestic public opinion.[47]

To clarify this aspect of Panarin's conceptualisation of information warfare, it is important to provide several additional definitions used by other proponents of this concept. Thus Professor Yuri Grigor'yev states that information wars are:

Non-violent actions directed to alter or destroy the unified informational domain of the adversary state. The purpose of information war is not the destruction of people, but an alteration of certain fragmented variables that dominate the informational domain of a considerable part of the citizens to a degree when these variables fall out of the unified informational domain of the native country, thus forcing these citizens to start organising themselves in different opposing structures.[48]

Another example is given by Professors Vladimir Lisichkin and Leonid Shelepin, who state that information wars are waged 'by a direct influence on the public consciousness, on the souls of people. The main purpose is to coerce masses to act in a desired direction, even against their own general interests, and in the adversary camp to split people and force them to rise one against another.'[49]

In other words, according to Panarin and other advocates of his theory, information warfare is primarily intended to subvert an adversary's political power by controlling and manipulating the informational trends that shape the actions of the elite in general, and public opinion in particular.

Panarin defines three main stages of information warfare. The first is a strategic political analysis that includes the 'collection, aggregation and exchange of information on adversaries and allies for the purpose of conducting active actions'. The second, informational influence, is based on the 'infiltration of negative commentaries and disinformation into the informational domain of the adversary, as well as the suppression of the adversary's attempts to get the information that he requires'. Finally, the third stage, informational defence, involves 'blocking the disinformation dispersed and infiltrated by the adversary.'[50] Thus, according to Panarin, information warfare targets the minds of the political elite and the general population to affect public opinion and thus influence the opposing side's political decision-making process.

Whereas Dugin defines different types of networks that play a key part in net-centric war, Panarin uses the term 'social object' to define a similar phenomenon. According to Panarin, social objects are the main targets in information warfare. Whether they are groups of individuals or social groups, they are 'relatively stable aggregates of people existing within the scope of historically defined society'. Panarin defines three different types of social objects: large, medium and small. The first, the 'large social groups', are groups of people on the state level, such as different social classes, professional groups and so forth. While these groups do not require direct contact between their members, they are responsible for the development of 'behavioural norms, spiritual-ethic values and traditions'. The 'medium social objects' are different groups of people with more defined associations, such as commercial and industrial institutions and organisations, military units, academic institutions, citizens of a city or neighbourhood. Though some of these groups might have clearly defined and controlled communication networks (e.g. banks, organisations, military units), some might not, as the aggregation of these groups is more spontaneous (e.g. a neighbourhood). The importance of the 'medium social groups', according to Panarin, is their role as a conductor of informational trends between the large social objects and the small ones, and vice versa. The last group is the 'small social objects', which Panarin defines as the 'aggregations of individuals, in which all members directly contact each other, such as family, small military unit, small business, company of neighbours or friends'. The information interactions within each group are structured according to the influence of the outside environment, as well as the impact of internal relations. It is important to understand this conceptual division, according to Panarin, as the processes of informational-psychological influence are defined not only by defined political goals 'but also by the objects of informational influence (large, medium and small social groups within one's country, or [within other] friendly, neutral or enemy states)'.[51]

Like Dugin, Panarin's ideas are ideologically informed by his opposition to the West; in his view, geopolitics has been dominated for several centuries by the struggle between two main civilisations—the sea-oriented (the British Empire and the United States) and the continent-oriented (Eurasia—Germany and Russia). Although this struggle has expressed itself many times in the form of physical clashes (i.e. wars), it has always been accompanied by information warfare—before, during and after the wars.[52] Moreover, Panarin claims that the West was able to master information warfare in the twentieth century, and it was this that ultimately led to the dissolution of the Soviet Union, since the 'main cause of the geopolitical catastrophe of 1991 was a defeat in the informational war, which lasted for 48 years'.[53] According to Panarin's interpretation of the Cold War, the main informational offensive of the United States was aimed at compromising and destabilising the Soviet political elite by targeting the weakest element of the Soviet political establishment—the transfer of power.[54] Referring to George Kennan's famous article 'The Sources of Soviet Conduct',[55] Panarin argues:

> Kennan succeeded in finding the weakest link in the U.S.S.R.—the problem of the smooth transfer of power from one individual person or group of people to another. Thus, it was Kennan, a person who lived in Russia for many years, who accurately defined the direction of the main attack in the information war against the U.S.S.R.[56]

Therefore, according to Panarin, the Soviet Union collapsed due to the general unprofessionalism of the Soviet political and military establishment, rather than the skilful exploitation of the informational domain by the United States. As he puts it:

> The Central Committee of the Communist Party and the KGB were acting formulaically; although the true intentions of the adversary in the informational-ideological struggle were revealed, adequate counter-measurements were not implemented. [As] nobody prepared experts on informational counter-measurements, the Soviet special agencies failed to prevent the dissolution of the U.S.S.R.[57]

Yet he also argues that this war is not over, as the struggle between the political elites of the West and Russia did not end in 1991. Since the information war continues to be 'the major tool of contemporary world politics, [and] the dominant way to achieve political and economic power in the 21st century', Russia, according to Panarin, continues to be targeted by Western political actors in the informational sphere.[58]

To avoid repeating the Soviet Union's defeat in the information war against the West, Panarin suggests that 'the existence of Russia depends on whether a new political elite will be formed—a passionate Russian political elite capable of an adequate response to the global challenges of the 21st century'. This new political elite, he maintains, should be based on intellectuals from the liberal arts and science, the senior leadership of the security services and military, together with representatives of big and medium capital. The strategic purpose of such an elite should be 'the formation of a positive global public opinion of Russia':

> Only a new Russian political elite, capable of skilfully conducting the geopolitical information confrontation, can create favourable conditions for the prosperity and development of the individual, society and the state, [and] to achieve its national and economic interests in the international arena.[59]

Panarin concedes that the formation of this new political elite, as well as the development of the required organisational infrastructure, would take a long period of time and would require significant investment. Yet he also suggests a very detailed plan for what he calls 'The Directions of the Informational Geopolitics of Russia'. First, he argues for the establishment of a 'meaningful system of management' that would coordinate activities of information war in the financial sphere and the field of foreign policy, 'including foreign political and economic propaganda'. These can be achieved by the coordination and integration of authorities, the mass media, security services of the biggest

Russian corporations and governmental information-intelligence agencies to 'organise defensive and offensive informational operations (on the territory of Russia and abroad)'.

The next step, according to Panarin, should be the organisation of a multifaceted exchange of information within a framework of private-governmental partnership (i.e. 'the establishment of a private-governmental intelligence agency to deal with the problems of informational war'). This agency would coordinate the efforts of different analytical institutions and inform the leaderships of the government and big corporations. It would also develop the 'conceptual, legislative, methodological and methodical base to wage informational warfare'. Together, Panarin concludes, these would not only allow Russia to successfully 'wage strategic informational struggle in the financial sphere and the field of foreign policy' but would also 'prepare the highest executive elite of the country (political, economic, information-ideological) to actively conduct information war in crisis situations (armed conflict, informational conflict, economic conflict, etc.)'.[60]

Like Dugin's concept of net-centric warfare, the essence of Panarin's information warfare is based on a narrative of Western aggression against Russia, which led the Russian people to destroy their country twice in the twentieth century (the fall of the Russian Empire in 1917 and the dissolution of the Soviet Union in 1991). As Panarin states: 'First and foremost, it [Russia] has to realise that for hundreds of years the informational-ideological sphere has been the weakest link of the Russian governance, thus leading the country to fall apart twice in the 20th century.'[61]

Moreover, similar to Dugin, who is also a supporter of Vladimir Putin and his efforts to counteract the Western net-centric war being waged against Russia, Panarin praises the actions of the president in countering the 'Western Information War against Russia'. Panarin argues that while Putin has become a major target of the information war against Russia following his declaration of a new Russian doctrine (which he claims is a 'doctrine of the Eurasian integration')[62] his successful return to office in 2012, after a very difficult electoral campaign, signified that: 'Russia's sober powers succeeded to block ... the repetition of the tragic events of February 1917 and 1991. Putin's victory means that Russia's people [successfully] defeated the foreign informational aggression.'[63]

This ideological theorisation of information warfare has proven popular within the Russian academic community. Since the mid-2000s, a vast number of articles and books have been written that have used Panarin's concepts and ideas to interpret the deteriorating relationship between Russia and the West, claiming that Russia has to defend itself against the informational offensive being waged against it by the West.[64]

Dugin's and Panarin's theories can be summarized by outlining their five shared aspects. First, both claim that the West (first the British Empire, then the United States) has been continuously and purposefully attempting to intervene and undermine the Russian political system before, during and after the Cold War (i.e. as part of an inter-civilisational struggle for dominance). Secondly, the main Western strategy has been an informational offensive that has targeted both Russian and international public opinion, manipulating the flow of information in political, diplomatic, financial-economic and military affairs. The third aspect is the claim that, in addition to the manipulation of information from the outside, Western strategy has aimed to create a 'fifth column' within Russia in an attempt to dismantle Russia from the inside. Fourth, as a response to these old and new threats, both scholars argue that Russia should nurture a new political elite to be sufficiently patriotic to overcome the 'Western Net-Centric/Information War', making Russia the political, cultural, economic and military centre of Eurasian civilisation. Finally, both praise Putin's achievements in counteracting the 'Western Net-Centric/Information War' by promoting the Eurasian project and nurturing a new elite able to defend itself, as well as transferring the battlefield of this war on to the West's territory.

5

THE RISE OF *GIBRIDNAYA VOYNA*

As with many other military concepts, Russia's interest in the theory of *gibridnaya voyna* (hybrid warfare) began with an observation of the West.[1] Russian military thinkers and analysts had already taken note of the US debate on hybrid warfare in 2009,[2] and in 2013 one of Hoffman's first articles was translated and republished in *Geoplotika* (Geopolitics), the journal of the faculty of social sciences at the Lomonosov Moscow State University.[3] Since then, Russian scholars have become increasingly interested in the concept, generating productive discussions in military and academic circles,[4] including high-profile conferences and seminars organised by the Military University of the Ministry of Defence of the Russian Federation, the Lomonosov Moscow State University and the Financial University under the Government of the Russian Federation, among others.[5]

Yet when analysing the works of Russian scholars, strategists and military thinkers, it quickly becomes clear that the only common ground between Hoffman's theory of hybrid warfare and *gibridnaya voyna* is the name. In interpreting hybrid warfare, the focus of Russian scholars and strategists has not been based on a 'blind' adaptation of the Western theory, but an attempt to reconceptualise it within the context of the Russian political–military experience and the Russian theoretical understanding of the phenomenon of war.[6]

As was discussed in the previous two chapters, since the mid-1990s the Russian conceptualisation of war as a socio-political phenomenon has been shaped by the defeat of the Soviet Union in the Cold War. In their analyses of the Cold War and the causes of the Soviet defeat, Russian strategists and political scientists have emphasised two main aspects in new conflicts: (1) the aim to break the spirit of the adversary's nation by a gradual erosion of its culture, values and self-esteem; and (2) an emphasis on political, informational (propaganda) and economic instruments, rather than on physical military force.[7]

Consequently, since the 2000s, inspired by the writings of Dugin, Panarin and other proponents of the same narratives,[8] the Russian academic and political discourse on net-centric and information warfare has flourished, with a vast number of books and articles being published on the subject, and even a special journal titled *Informatsionnyye Voyny* (*Informational wars*) has been established under the supervision of the famous General Makhmut Gareev, the head of the scientific-editorial council.[9]

In tandem with this discourse, led mainly by political scientists, there has been a renewed academic interest in Messner's concept of subversion-war. Messner's works have been widely republished since the 1990s as separate books or as chapters in edited compilations,[10] with his theory of subversion-war becoming an important prism through which to examine contemporary political, military and economic confrontations.[11]

It is important to note that the proponents of Messner's subversion-war, as well as Dugin and Panarin and their theories of net-centric and information warfare, have all been conceptualising the very same idea—corrupting and undermining the adversary internally to achieve certain political goals (i.e. global domination) without escalating a given conflict to a direct physical military confrontation. Therefore, these independent, but similar theories had successfully coexisted in Russian academic and analytical discourse up until the Ukrainian Crisis, after which the term *gibridnaya voyna* gained widespread popularity among Russian scholars, political analysts and commentators.

In analysing the scope of literature produced in the last few years by Russian scholars on the phenomenon of *gibridnaya voyna*, it becomes clear that, while the US concept of hybrid warfare focuses chiefly on tactical military and operational activities 'directed and coordinated within the main battlespace to achieve synergistic effects',[12] the Russian theory is completely different in that it revolves around more abstract ideas and 'involves all spheres of public life: politics, economy, social development, culture'.[13] In other words, while Hoffman's theory of hybrid warfare argues for the amalgamation of regular and irregular forces and incorporates mixed operational and tactical methods, the concept of *gibridnaya voyna* emphasises a more abstract battlefield, on which actors fight to erode their adversaries' socio-cultural cohesion and protect their own.

THE RISE OF *GIBRIDNAYA VOYNA*

It is possible to identify three interconnected ways in which Russian scholars, strategists and military thinkers have approached these ideas. The first is a broad, philosophical approach. Traditional Russian military thought has always been inclined to conceptualise the phenomenon of war as a socio-cultural struggle rather than a clash of materialistic interests.[14] In the words of an 1899 Imperial Nicholas Military Academy (the General Staff Academy) manual on strategy: 'The history of nations offers a change between cultures as an outcome of wars. A war is a struggle for existence ... War is this special way, by which one culture, the stronger one, overwhelms the other weaker one.'[15]

The proponents of this approach see nothing new in the concept of *gibridnaya voyna* as, from their perspective, a war has always been 'a confrontation between two or more autonomous groups ... in order to improve their financial, social, political or psychological condition, or [their] overall survival chances'[16]—as such, this concept is merely a new term for a very old phenomenon.[17]

The second approach to the idea of *gibridnaya voyna* constructs it as a phenomenon that explicitly characterises contemporary history with the aim of trying to preserve the broad philosophical perspective of Russia's traditional way of military thinking. Though Hoffman originally defined hybrid war as a hybrid of regular and irregular tactics, formations and technologies,[18] Russian scholars and military thinkers have sought to broaden the concept. In their view, *gibridnaya voyna* is the 'creation of external controlling mechanisms, an infiltration of subversive and destructive concepts, projects and programs, a formation of an agency of influence and promoting its representatives to power.'[19]

The proponents of this approach claim that *gibridnaya voyna*, unlike hybrid warfare, is not limited to military activity on the battlefield. According to them, the main purpose of this type of war is to avoid the traditional battlefield and destroy the adversary via a hybrid of ideological, informational, financial, political and economic methods that dismantle the socio-cultural fabric of society, leading to its internal collapse (i.e. a 'Colour Revolution').[20] Thus, according to this group of Russian scholars and strategists, the aims of *gibridnaya voyna* are:

> A full or partial disintegration of the [adversary] state, a significant alternation of the direction of its internal or foreign policy, a replacement of the state's leadership with a loyal regime, an establishment of foreign ideological and finance-economical control over the state, its chaotisation and subordination to the dictats of foreign states [that wage hybrid war against it].[21]

The third way in which the conceptualisation of *gibridnaya voyna* has been approached in Russia is by explicitly focusing on one dimension—informational warfare. The proponents of this approach argue that the informational struggle is the essence of *gibridnaya voyna*, as this type of war requires:

> Creating a necessary socio-political pseudo-reality in the media-space, which generates the required models that design people's perception of 'enemies' and 'friends', 'aggressors' and 'victims', 'invaders' and 'freedom fighter' ... aiming to mobilise the active [part of the] population of [adversary] countries to generate by themselves military and political actions without the direct and apparent participation of the true beneficiaries of 'gibridnaya voyna'.[22]

According to this view of *gibridnaya voyna*, information warfare is a primary component of this type of war[23] as 'information wars are the core of *gibridnaya voyna*'.[24] Moreover, Panarin, whose ideas on the 'Western information war against Russia' were discussed in the previous chapter, simply equates *gibridnaya voyna* with information war, arguing that:

> In my opinion, the fall of the powerful state [the Soviet Union] occurred as a result of a systematic hybrid war waged by the West against the U.S.S.R., though the term itself appeared later ... The term 'hybrid' means the simultaneous usage of several types of pressure (information-ideological, finance-economical, etc.) on an adversary, where the actions of armed forces play an important, but only secondary, part.[25]

To understand why the Russian concept of *gibridnaya voyna* bears a closer resemblance to the theories of subversion, net-centric and information wars than Hoffman's concept of hybrid warfare, it is important to examine the specific context of the 'Russification' of Hoffman's theory, focusing on two main aspects: (1) the reason for its appearance; and (2) the remarkable difference between the original US conceptualisation of hybrid warfare and the Russian concept of *gibridnaya voyna*. The first aspect has been explained by Professor Pavel Afanas'yevich Tsygankov, the acting head of the faculty of political science at the Lomonosov Moscow State University. In 2015, Tsygankov argued that due to the Western discourse surrounding the 'so called "hybrid warfare", which Russia allegedly wages in Ukraine' it seems 'reasonable to analyse the definition of "hybrid warfare" in the context of traditional and alternative approaches to the research of wars and armed conflicts'.[26] Hence the rise of the term *gibridnaya voyna* in Russia can be seen as a direct response by the Russian academic community to the politicisation of this term in the West within the context of the Ukrainian Crisis.

The connection (or lack thereof) between hybrid warfare and *gibridnaya voyna*, on the other hand, has been explained by Konstantin Valentinovich

THE RISE OF *GIBRIDNAYA VOYNA*

Sivkov, captain 1st rank (ret.), doctor of military science and the founder of the Academy of Geopolitical Problems (an independent military think-tank). Answering the question of the nature of hybrid warfare, Sivkov states that 'the term itself was introduced by foreign military experts' and 'we use this term only because it has caught on'. However, when he then goes on to define *gibridnaya voyna*, he used the narratives espoused by the proponents of the theories of Messner, Dugin and Panarin, rather than Hoffman:

> In contrast to the idea of traditional war, which underlines the destruction of an enemy's regular forces and the following enforcement of peace by deposing the existing authority from power ... the idea of hybrid war is based on ... first, the disposing of the acting authority, and then the destruction of the military capacity, security system and economy by the establishment of puppet authorities.[27]

The conceptual and analytical literature produced in Russia on the topic of *gibridnaya voyna* can easily be divided into two main groups. The first does not mention the US origins of the term, instead utilising Russian interpretations of the geopolitical realities based on the narratives put forward by Messner, Dugin, Panarin and their followers.[28] The second group does mention the US origin of the concept of hybrid warfare, but these authors emphasise the subversive nature of *gibridnaya voyna* rather than the ideas in the original concept of hybrid warfare.[29]

Regardless of whether the theory of *gibridnaya voyna* has been independently conceptualised by Russian scholars and military thinkers, or whether it is based on the interpretation of the theory of hybrid warfare within the context of Russian national political and military thought, hybrid warfare and *gibridnaya voyna* are two completely different things. Whereas hybrid warfare represents the complexity of military threats in the twenty-first century, based on a mixture of regular and irregular tactics, technologies and capabilities, *gibridnaya voyna* focuses on ways that political actors can undermine their adversaries by eroding their domestic and international political legitimacy and stability.

While *gibridnaya voyna* is conceptually closer to the theories of subversion-war, net-centric or information wars than to the original definition of hybrid warfare, Russian strategists have found the Western (US) origins of the term 'hybrid warfare' to be very useful: it allows them to claim that the United States invented this type of warfare and has used it successfully (against Russia and its allies) for several decades. In other words, Russian analyses of 'US hybrid wars' in Latin America in the second part of the twentieth century,[30] as well as more contemporary cases of 'US hybrid warfare' in the Middle East,[31] or, indeed, the 'US hybrid war' in Ukraine,[32] are conceptually closer to the works of Messner, Panarin and Dugin than to Hoffman.

Another interesting aspect of the rise of the term *gibridnaya voyna* in Russian discourse is the reluctance of members of the Russian military, like their US counterparts,[33] to adopt this term. When discussing the changing nature of war in 2013 and 2014, the Russian chief of the General Staff, General Valery Gerasimov, preferred to describe the new developments as a 'new adaptive approach to the use of military force'[34] rather than *gibridnaya voyna*, despite his explicit articulation of its narratives:

> In the 21st century there is a visible tendency of erasing distinctions between a state of war and peace. Wars are no longer declared, but once started—[they] do not follow the pattern that has been usual for us ... The emphasis of the used methods of confrontation is shifting towards the widespread use of political, economic, information, humanitarian and other non-military measures implemented by taking an advantage of the protest potential of the population. All these are complemented by military means of a hidden character, including the implementation of information warfare actions and special operations forces.[35]

Thus, to summarise the rise of the concept of *gibridnaya voyna* in Russian academic, political and professional discourse, it is important to emphasise its six main characteristics. The first is that, despite sharing the same name, the concept of *gibridnaya voyna* differs fundamentally from the original theory of hybrid warfare as conceptualised by Hoffman and his followers in US military circles. Whereas Hoffman's theory highlights the hybridisation between different military means, tactics and technologies, the Russian theory of *gibridnaya voyna* emphasises the links between military and non-military means of political struggle, where the main roles are reserved for non-military means, and military means play a supporting role (if they play any role at all). This difference ultimately leads to the second characteristic of *gibridnaya voyna*—the means by which it is fought. Unlike more traditional concepts of war (regular and irregular, conventional and unconventional), the basis of which is to destroy the political power of an adversary by fighting against its physical might (i.e. armed forces), the aim underlying the theory of *gibridnaya voyna* is to destroy the political cohesion of an adversary from the inside by employing a carefully crafted hybrid of non-military means and methods that amplify political, ideological, economic and other social polarisations within an adversary's society, thus leading to its internal collapse. In this sense, *gibridnaya voyna* shares more with the US concept of 'political warfare' than hybrid warfare:

> Political war is the use of political means to compel an opponent to do one's will ... [and it] may be combined with violence, economic pressure, subversion, and diplomacy, but its chief aspect is the use of words, images, and ideas, commonly known, according to context, as propaganda and psychological warfare.[36]

The third important attribute of the theory of *gibridnaya voyna* is that it does not add any conceptual novelty, but simply repeats and paraphrases ideas that have been at the centre of academic and professional discourse since the mid-1990s. As discussed above, the proponents of the theory of *gibridnaya voyna* simply replicate the narratives expressed by pre-existing concepts of subversion-war, net-centric or information wars, or even equate them. Yet the presence of these narratives at the centre of academic, political and military discourse before the arrival of *gibridnaya voyna* has obviously reduced traditional institutional criticism of new concepts, facilitating the rising popularity of *gibridnaya voyna*. Thus, with its arrival, the concept of *gibridnaya voyna* found fruitful conceptual ground that had been fertilised by previous theories for almost two decades. This leads to the fourth characteristic—the reason for the appearance of the term *gibridnaya voyna*. Although the ideas expressed by *gibridnaya voyna* are not new, the term 'hybrid warfare' was 'Russified' for political reasons, rather than due to the conceptual limitations of previous theories. First, as discussed, the introduction of the term was intended to mirror the reaction of Western analysts, scholars and politicians, who accused Russia of implementing hybrid warfare during the Ukrainian Crisis of 2014–15. Secondly, by adapting Western terminology, Russian proponents of *gibridnaya voyna* were able to argue that the West had already mastered its techniques and methods.

This leads directly to the fifth characteristic. According to Russian scholars, *gibridnaya voyna* is a method of a geopolitical confrontation explicitly used by the West (mainly the United States) against its known or potential adversaries, first and foremost Russia, in situations when open war is impossible or undesirable. As has been mentioned, in their analyses and conceptualisations of the defeat of the Soviet Union in the Cold War, the proponents of the conceptual narratives that constitute *gibridnaya voyna* argue that the USSR fell victim to Western methods and techniques based on political, economic, ideological and other non-military types of subversion. Since Russia is the direct successor of the Soviet Union, and due to the Kremlin's efforts (at least under President Putin) to resist the Western 'offensive', Russia has remained the main target of the US *gibridnaya voyna*.

The sixth and final important feature of the concept is that it has not been officially adopted by the Russian military. As discussed above, Russian military thought has traditionally tended to conceptualise the phenomenon of war within a broader framework of socio-cultural and political confrontations, stipulating that war is not only an instrumental clash of armed forces for political goals but also involves other dimensions, such as the political, ideological, informational and economic dimensions. Therefore, it might be argued that the concept of *gibridnaya voyna* is more than familiar to the Russian military leadership, and it is not surprising that in 2016 General Gerasimov finally used this term when referring to the changing nature of the conflict:

> In contemporary conflicts, the emphasis of the methods of confrontation is more frequently shifting towards an integrated application of political, economic, informational and other non-military measures, implemented with the support of the military force. These are so-called hybrid methods. Their purpose is to achieve political goals with a minimal military influence on the enemy ... by undermining its military and economic potential by information and psychological pressure, the active support of the internal opposition, partisan and subversive methods ... A state that falls under the influence of a hybrid of aggression usually descends into a state of complete chaos, political crisis and economic collapse.[37]

However, it is important to remember that Gerasimov's articles were published in *Voyenno-Promyshlennyy Kurier* (*Military-Industrial Courier*), a private newspaper owned by the quasi-government-controlled Almaz-Antey company, and not in *Voennaya Mysl'* (*Military Thought*), published by the Russian General Staff, the main platform for peer-reviewed military/academic discussion. Thus, as one US analyst put it, 'it is important to keep in mind that Gerasimov is simply explaining his views of the operational environment and the nature of future war, and not proposing a new Russian way of warfare or military doctrine.'[38] Hence despite Gerasimov's personal opinion on the theory of *gibridnaya voyna*, expressed for a specific audience, which 'may not even be in the Russian armed forces, but instead in Russia's senior political leadership',[39] and despite explicit calls from academics for the military to embrace this concept in official publications and terminology,[40] the term *gibridnaya voyna* appears neither in the *Dictionary of Military Terms* (*Voyennyy entsiklopedicheskiy slovar'*) published by the Russian Ministry of Defence,[41] nor in any official military doctrinal publication. In other words, like its conceptually related predecessors, the theory of *gibridnaya voyna* seems to include too many ambiguous non-military aspects to answer the practical questions that General Gerasimov and his General Staff need answers for: 'what is contemporary war, what should the military be prepared for, [and] what should it be armed with?'[42]

PART 3

THE POLITICISATION OF HYBRID WAR AND *GIBRIDNAYA VOYNA*

6

'THE RUSSIANS ARE COMING'

THE POLITICISATION OF RUSSIAN HYBRID WARFARE

As discussed in Part 1, the idea of hybrid warfare was developed by the US military during the 2000s as an operational concept intended to describe the changing nature of contemporary conflicts. By the beginning of the 2010s, this concept had gained as much popularity as criticism. As an outcome of both, there were different attempts to interpret Hoffman's ideas, the most methodological of which was published by NATO in 2012. NATO's initial enthusiasm for the idea of hybridity was not shared by its member states, yet this changed with the Ukrainian Crisis, which brought the idea back to the centre of academic, military and political discourse. This chapter seeks to analyse this discourse. The first part of the chapter focuses on the main criticism of the initial US concept of hybrid warfare and its later reconceptualisation, which was intended to answer the disadvantages and weaknesses outlined in the critiques of the concept. The second part discusses the revival of hybrid warfare in the context of Russia's actions in Crimea and eastern Ukraine and the rise of the narrative of Russian hybrid warfare. The final part of the chapter analyses the politicisation of this narrative in the West, focusing on the ways in which the conceptual debate on contemporary warfare has been reshaped to serve different political purposes.

RUSSIAN 'HYBRID WARFARE'

The critiques of the concept of hybrid warfare

As discussed in previous chapters, the concept of hybrid warfare was the product of military thinking in the United States. Hoffman and other proponents of this theory based their conceptualisation on the observation of the Israeli experience in the Second Lebanon War, as well as other military conflicts during the twentieth century.[1] After several years of intense discourse within the US military community (mainly the US Marine Corps), the idea of hybrid wars and hybrid threats started to attract broader attention, leaving the cocoon of US military thought and exposing itself to criticism and independent interpretations.

It is possible to point to three main characteristics of this theory that have been targeted most frequently by its critics: the alleged novelty of the described phenomenon; the ambiguous nature of the concept; and its 'a-strategic' nature. The first originated with military historians, who have claimed that there is nothing new in the idea of hybridity as proposed by Hoffman, and who have criticised the 'intellectual apparatus of the American military' for its failure 'to understand the future by reference not only to the distant past, but also to the immediate past as well'.[2] For example, Peter Mansoor claims that while 'some defence analysts have posited the emergence of a new type of war—hybrid war', a careful examination of history shows that 'there is little new in hybrid war as a concept'.[3] Similarly, Max Boot states that 'hybrid warfare is a modern term for an ancient practice', the origins of which 'can be traced to sometime in the last 5,000 years, before which there were no conventional armies'.[4] In a similar vein, Williamson Murray argues that: 'Despite the surprise that the events in Lebanon elicited in the American defence community, the historical record suggests that hybrid warfare in one form or another may well be the norm for human conflict, rather than exception.'[5]

The second criticism of the concept of hybrid warfare comes from the community of military strategists, who claimed that Hoffman's concept is too ambiguous to be used in practical situations. For example, Dan Cox, together with several co-authors, argues that:

> The concept of hybrid threats (or hybrid warfare), as defined by its main proponents, is indeed unclear, incomplete, and often unhelpful. [As] by arguing that individual units (or even separate but aligned units) can somehow simultaneously (or easily and quickly) switch back and forth between conventional, irregular, and criminal activities elevates the enemy to mystical status.[6]

According to Hew Strachan, 'it is unclear whether "hybrid wars" are those which occupy some middle point in the spectrum between regular and irregular, or whether they are characterised by simultaneous activity on both ends of that spectrum'.[7] Moreover, in the words of Sibylle Scheipers, 'owning to its internal inconsistency, the concept of hybrid warfare seems to be difficult to operationalise into any form of purposeful action' as 'by lighting the alleged effectiveness of a "strategy" that leaves no room for defining its own parameters of success'.[8]

This leads directly to the third field of criticism of the concept of hybrid warfare—its 'a-strategic' nature. According to some strategists and military thinkers, 'the fundamental problem with the hybrid warfare analysis is that it ignores the role of interaction in strategy'.[9] As was discussed in Chapter 2, the concept of hybrid warfare was developed by and for the US military, with the aim of answering two basic demands of US military culture—pure military orientation and practicality.[10] This approach helped the concept to gain in popularity within the US military establishment, yet it also attracted significant criticism. For example, whereas Bettina Renz argues that hybrid warfare does not represent strategy, but merely an operational approach,[11] Robert Mihara claims that: '[Hybrid] threat-based approach makes eminent sense in prioritizing initiatives for developing operational doctrine or in campaign planning, but it makes far less sense when promulgating a strategic plan for an Army institution that is posturing itself for the long term.'[12]

Brett Friedman takes an even more critical stance with the claim that by conceptualising hybrid warfare 'we are simply discussing warfare—professionally and historically known as tactics', and concluding that:

> Forays into the concept *du jour* only distract from the underlying strategic principles that truly are different from conflict to conflict, combatant to combatant. Facile changes in tactical trends do not constitute revolutions in theory. Hybrid warfare is perhaps the most egregious example of this. Rather than a new threat or form of warfare, hybrid warfare is just a misrepresented and misunderstood recognition of the utter lack of a fundamental difference between conventional and irregular warfare at the tactical level.[13]

Furthermore, Antulio J. Echevarria argues that by focusing on what contemporary wars are (or are not) the concept of hybrid warfare is actually counterproductive to strategic decision-making and planning.[14]

Hybrid warfare is reinvented

Either as a result of the described criticism of the a-strategic nature of Hoffman's theory and its narrow (operational, or even, tactical) perspective on the nature of contemporary conflicts, or as a part of the natural development of any voguish concept, the theory of hybrid warfare was reconceptualised, encompassing additional dimensions that were lacking in the original concept. As Lawrence Freedman puts it: 'as with many similar concepts, [hybrid warfare] once adopted as a term of art has tended towards a wider definition'.[15] The first attempt to offer an alternative definition of hybrid warfare that would take it out of its purely military niche, where it was placed by those who originally formulated it, was probably made by David Kilcullen. As early as 2009 (Hoffman's first major publication was in 2007), Kilcullen stated that 'today conflicts clearly combine new actors with new technology and new or transfigured ways of war', and argued that hybrid warfare is a phenomenon that combines different military and non-military actions as a result of governments', as well as non-government organisations', desires to achieve internal and international influence with or without applying direct or indirect violence.[16]

The first comprehensive and methodological attempt to reconceptualise hybrid warfare and hybrid threats and elevate these concepts to the level of strategy was published by NATO in 2010 in its *Bi-Strategic Command Capstone Concept*, which states that:

> Hybrid threats are those posed by adversaries, with the ability to simultaneously employ conventional and non-conventional means adaptively in pursuit of their objectives ... Hybrid threats are comprised of, and operate across, multiple systems/subsystems (including economic/financial, legal, political, social and military/security) simultaneously.[17]

It is not surprising that NATO was the first organisation to expand Hoffman's tactical–operational understanding of hybrid warfare into a strategy. Since NATO is not a purely military organisation and involves a significant political component, Hoffman's military conceptualisation of hybridity did not suit the way it thinks about conflicts.

The task of developing and articulating this new conceptual approach to hybrid threats was assigned to NATO Allied Command Transformation (ACT), which, supported by the US Joint Forces Command Joint Irregular Warfare Centre (USJFCOM JIWC) and the US National Defense University (NDU), conducted a series of specialised workshops and experiments to identify potential hybrid threats and examine viable and effective strategies to meet them.[18] As an outcome of this collaboration, NATO arrived at the following understanding of hybrid threats:

> Admittedly, hybrid threat is an umbrella term encompassing a wide variety of existing adverse circumstances and actions, such as terrorism, migration, piracy,

corruption, ethnic conflict, and so forth. What is new, however, is the possibility of NATO facing the adaptive and systematic use of such means singularly and in combination by adversaries in pursuit of long-term political objectives, as opposed to their more random occurrence, driven by coincidental factors.[19]

USJFCOM JIWC defined these threats as being posed by 'Multi-dimensional adversaries [that] employ a complex blend of means that includes the orchestration of diplomacy, political interaction, humanitarian aid, social pressures, economic development, savvy use of the media, and military force.'[20]

Despite the productive debate generated by the concept, the political unwillingness of NATO's members to invest resources in developing the capabilities required to meet these threats led NATO to switch its central focus from hybrid threats in 2012, even though it encouraged its member states and NATO Excellence Centres to continue working on their own.[21]

This encouragement did not fall on completely deaf ears. In 2012, for example, the Swedish National Defence College conducted its own analysis in an attempt to understand the relevance of hybrid threats to Sweden's security. The examined scenario included an imaginary enemy that simultaneously employs a mix of traditional threats (such as infiltration of Special Forces into Swedish territory and an attempt to sink a hijacked oil tanker in the middle of a sensitive maritime environment zone) and non-traditional ones (such as a cyber-attack aimed to deface governmental sites or destroy sensitive infrastructure—e.g. a nuclear power station—and threats to hack the pacemaker of a high-ranking Swedish official).[22] Such an interpretation of hybrid threats obviously stretched the definition to an extreme, and it is not surprising that Sascha-Dominik Bachmann and Håkan Gunneriusson, who reported and interpreted this example in a series of articles, stated that the umbrella of hybrid threats extends to:

> Multimodal, low-intensity, kinetic as well as non-kinetic threats to international peace and security [that] include cyber war, asymmetric conflict scenarios, global terrorism, piracy, transnational organised crime, demographic challenges, resources security, retrenchment from globalisation and the proliferation of weapons of mass destruction.[23]

This example also serves to demonstrate the slow transformation of the concept of hybridity, from a purely military dimension, as was initially proposed by Hoffman and other proponents in the US military, to an all-embracing idea that encompasses the whole spectrum of possible military and non-military threats that endanger state security and internal and external stability. Moreover, this preoccupation with the theory of hybrid warfare, beyond NATO supervision, meant that it was not only possible to widen its conceptualisation without significant criticism but also to maintain the continuity of the theoretical debate, preparing the ground for the comeback of this concept after the beginning of the Ukrainian Crisis in 2014.

Many researchers have outlined the fact that analyses and commentaries on the concept of hybrid warfare have proliferated since the beginning of the Ukrainian Crisis in 2014.[24] This process is discussed in detail in the following parts of this chapter; here, it is simply important to underline the conceptual transformation that has been occurring with the theory of hybrid warfare within the context of this debate.

Analysing 'the politico-military methods employed by Russia' in Crimea and eastern Ukraine, which are 'generally labelled [as] "hybrid" warfare', the International Institute for Strategic Studies (IISS) concludes that hybrid warfare includes:

> the use of military and non-military tools in an integrated campaign designed to achieve surprise, seize the initiative and gain psychological as well as physical advantages utilising diplomatic means; sophisticated and rapid information, electronic and cyber operations; covert and occasionally overt military and intelligence action; and economic pressure.[25]

While this conceptual extension of hybrid warfare to any possible hostile activity was titled by IISS as a 'rediscovery', it can hardly be described as such. Instead, this all-embracing definition is a direct outcome of the above-mentioned discussions that have slowly extended the umbrella of hybrid warfare, starting with NATO's approach to hybrid threats in 2010 and the subsequent developments in different European schools of military thought (e.g. the Swedish National Defence College). In this way, and since the events of 2014, the concept of hybrid warfare, in the context of Russian actions in Ukraine and elsewhere, has been at the centre of Western (mainly European) military conceptual discourse. The NATO Defence College (NDC) was the main driving force behind this discourse, as, for example, in 2015 more than half of all research publications produced by NDC discussed so-called 'Russian Hybrid Warfare'.[26]

In analysing the conceptual understanding of hybrid warfare as it has developed among NATO officials and experts in the context of their interpretation of the Kremlin's actions in Ukraine, it is possible to point to a general tendency to extend the definition of this type of warfare to all possible combinations of military and non-military means. For example, in 2015, General Philip M. Breedlove, the Supreme Allied Commander Europe, stated that 'what we now commonly refer to as hybrid war' is a 'continuum of threat, including unconventional and conventional methods'.[27] Diego A. Ruiz Palmer, an expert from the International Staff at NATO Headquarters in Brussels, explained this extended vision of NATO's contemporary interpretation of hybrid warfare: 'In effect, hybrid warfare bridges the divide between the hard and soft power applications that result from the technological and informational revolutions of the last three decades in ways that maximise asymmetric advantages ... as well as minimise risks and costs.'[28]

As the following parts of this chapter discuss the rise of Russian hybrid warfare as the main label for Russian actions in Ukraine, it is important to highlight the conceptual context within which this labelling has occurred. As discussed previously, the concept of hybrid warfare was originally developed and promoted within the US military as an operational approach, intended to improve the effectiveness of military units on the battlespace. Since then, the term has been significantly developed and transformed to the extent that the only mutual ground between the original concept of hybrid warfare and the rediscovered one is their title. Whereas the original concept describes a mixture of conventional and unconventional tactics, capabilities and technologies on the same battlespace, the more recent reformulation refers to a combination of hard and soft power in confrontations between two rival political actors. In other words, where the original concept was limited to the military realm, the more recent versions embrace the whole spectrum of possible military and non-military means used by an adversary to achieve its desired political goals. And as this interpretation was developed in light of the Russian reactions to the Ukrainian Crisis, it was the main conceptual context behind the rise of 'Russian hybrid warfare'.

The rise of Russian hybrid warfare

The Russian reaction to the Ukrainian Crisis in Crimea and eastern Ukraine undoubtedly surprised the Western community in general and NATO in particular. In January 2014, Heidi Reisinger from the NATO Defence College assessed that Russian military forces are 'neither a threat, nor a partner', claiming that:

> Many years of continual reform, underfunding, and the devastating effects of demographic trends have led the Russian armed forces to a situation where even senior military personnel raises doubts about the ability to provide national defence without tactical nuclear weapons ... All this makes Russia's military capabilities less efficient and hardly interoperable.[29]

Yet, just ten months later, in October of the same year, Reisinger's assessment had changed dramatically:

> Russia's recent behaviour and actions are often referred to as 'Hybrid Warfare'. They have been an effective and sometimes surprising mix of military and non-military, conventional and irregular components, and can include all kinds of instruments such as cyber and information operations. None of the single components is new; it is the combination and orchestration of different actions that achieves a surprise effect and creates ambiguity, making an adequate reaction extremely difficult, especially for multinational organisations that operate on the principle of consensus.[30]

While this change of attitude towards Russian military capabilities clearly demonstrates that NATO had been caught off-guard by Russia's performance in Crimea and eastern Ukraine, it also means that it is possible to identify the time period in which Russian actions began being described as hybrid warfare. In other words, to understand why and how 'Russia's recent behaviour' had started to be 'referred to as "Hybrid Warfare"', it is important to analyse the professional discourse between January and November 2014 in order to trace how and by whom this conceptual link was established.

An analysis of the political and military discourse on Russian activities in Crimea and eastern Ukraine, in the spring and early summer of 2014, demonstrates that Western experts were surprised not only by the effectiveness of the Russian military's performance but also by its successful coordination of different simultaneous activities in different dimensions: military (covert and overt), political, informational, economic and others. In the words of Marcin Zaborowski, the former director of the Polish Institute of International Affairs: 'I think the Russians have been very smart. Frankly, I think they have outsmarted us. They use commandoes and they pretend they are not Russian. In terms of information warfare, they have been extremely good.'[31]

The former US ambassador to NATO, Kurt Volker, concluded that: 'Russia is going to use special operations and intelligence forces, economic pressure, energy pressure, cyber-attacks and potential conventional force directly to achieve imperial goals.'[32]

Although many Western experts clearly observed the multidimensionality of Russia's activities, it was a former NATO adviser on security, Dutch Major General Frank van Kappen, who was among the first to create a conceptual link between Russian actions and the theory of hybrid war and hybrid threats, stating on 26 April 2014 that 'Russian actions in Ukraine are an example of Hybrid War.'[33] A broader connection was then made two months later, in late

June 2014, at the NATO Headquarters in Brussels. This conceptual breakthrough was made public by the then NATO Secretary General Anders Fogh Rasmussen following his meeting with President Obama in Washington in early July[34] (in fact, already in June, Rasmussen had argued, during a conference in London, that Russian actions were 'a manifestation of hybrid war').[35] This conceptual link was further cemented by a short documentary published by *NATO Review Magazine* on 1 July 2014. Titled 'Hybrid War—Hybrid Response?', the video started with the following words: 'At one point during the Ukrainian crisis, Russia had 40,000 troops lined up on the Ukrainian border, but when it came to sowing instability in Ukraine, it was not conventional forces who were used, but rather unorthodox and varied techniques, which have been dubbed hybrid warfare.'[36]

At this early stage of the debate, descriptions of Russian actions as hybrid war were designed to capture the mix of military and non-military means and methods employed by the Kremlin in Crimea and eastern Ukraine, rather than as a reference to the theory of hybrid warfare. Indeed, Paul King, the editor of *NATO Review Magazine*, maintained that when he titled the video 'Hybrid War—Hybrid Response?' he 'was not thinking of the NATO hybrid concept ... it just seemed an appropriate description'.[37]

This conceptual connection was made by Hoffman in an article published on 28 July 2014 on the *War on the Rocks* website. In his analysis of the increasing linkage between Russian actions and the concept of hybrid warfare, Hoffman made two very important points. The first was a criticism of his own original conceptualisation for being overly focused on combinations of tactics associated with violence and warfare, and not addressing 'instruments including economic and financial acts, subversive political acts ... or information operations'.[38] The second contribution Hoffman made at this stage of the debate was his suggestion of the need to stop thinking about wars in 'neat orthodoxies and Clausewitzian models about how "real wars" are fought and won', as well as his encouragement to conceptualise Russian actions as hybrid warfare, as 'perhaps [they are] better captured by hybrid, rather than political or "new"'.[39]

These steps were undoubtedly important in shaping the discourse on Russian actions in Crimea and eastern Ukraine as a hybrid war. However, it was after the 2014 NATO Summit in Wales that this association began to gain real traction. On the one hand, in its discussion of 'Russia's aggressive actions against Ukraine',[40] the Wales Summit Declaration did not refer directly to Russia's behaviour as hybrid war. Yet the idea of hybrid threats, which was initially dismissed in 2012, was clearly back in vogue, as NATO's members stated:

> We will ensure that NATO is able to effectively address the specific challenges posed by hybrid warfare threats, where a wide range of overt and covert military, paramilitary, and civilian measures are employed in a highly-integrated design. It is essential that the Alliance possesses the necessary tools and procedures required to deter and respond effectively to hybrid warfare threats ... We have tasked the work on hybrid warfare to be reviewed alongside the implementation of the Readiness Action Plan.[41]

The missing link between Russia and this concept of hybrid threats was provided at an Atlantic Council event on the first day of the summit by General Philip M. Breedlove, former Supreme Allied Commander Europe: 'What we see in Russia now, in this hybrid approach to war, is to use all the tools they have ... to stir up problems they can then begin to exploit through their military tool.'[42]

2014—the first wave of discourse on Russian hybrid war

The academic and professional discourse on Russian hybrid war after the Wales Summit in September 2014 can be divided into three different 'waves'. The first came as early as the autumn of 2014, and it was largely led by military thinkers from Eastern Europe.[43] It is possible to point to three principal narratives that appear in most of the articles written by these authors. The first is the reference to the video mentioned above, published by *NATO Review Magazine*, as proof of NATO's recognition that Russian actions in Ukraine represented a case of hybrid war.[44] While the video did not intend to draw a connection between the concept of hybrid threats and Russian actions, it seems that the authors of these articles understood it differently, using it as a seal of approval for this conceptual link. In the words of one of the articles: 'NATO has already named Russian actions as hybrid war ... On the NATO website, we find a short documentary movie ... where officials ... are trying to describe and find measures against this kind of threats.'[45]

This leads to the second shared characteristic of these articles—the conceptual linkage to the previously established concepts of hybridity. Rooting their analysis in the conceptual literature on hybrid warfare and hybrid threats, either in its limited formula, as put forward by Hoffman and his followers,[46] or in its broader understanding, as formulated by NATO in 2010–12,[47] these articles established a conceptual link between the theory of hybrid warfare and Russian actions. In other words, what was initially titled as 'Russian hybrid war', because 'it just seemed an appropriate description',[48] had become part of a much wider, almost-decade-old conceptual debate. As one Romanian expert concluded:

> Although NATO members initially did not know where to fit the Ukrainian conflict because of their national doctrines that separate conflicts in conventional and unconventional, the military studies prepared by experts in this field were able to impose ... [the concept of] *hybrid war* that is known in this moment by everyone.[49]

The third narrative that can be traced in most of the articles published by the East European experts in the second part of 2014 is a call for help from NATO members to respond to: 'The action of Russian Federation taken in Ukraine, in which manipulation, persuasion, conventional actions, unconventional actions, economic war, the energetic war, informational warfare including cyber warfare have been raised at the level of art.'[50]

And therefore: 'The democratic forces represented here by the North Atlantic Alliance, once [they have] identified the danger represented by the Russian Federation through its expansionary policy that it promotes in this moment and that has not changed over the centuries, would create the capabilities to counter these threats.'[51]

In other words, as a trio of Romanian officers put it: 'NATO has the responsibility to help their allies from the eastern border to defend against any aggression from Russia, especially those who are affected by the use of unconventional tactics like in Ukraine that may cause special challenges and threats to them.'[52]

2015: the second wave of discourse on Russian hybrid war

The second wave of literature to discuss Russian hybrid warfare as a major threat to the security of NATO's members began to appear in 2015, following the publication of the above-mentioned articles written by East European specialists.[53] Work on the subject produced in that year includes analysis from the Scandinavian[54] and Baltic states,[55] the UK,[56] the United States[57] and other Western countries, with NATO leading the discourse. In addition to a vast amount of research produced by NATO experts in 2015 on the topic of Russian hybrid warfare,[58] NATO also organised several high-profile conferences on the topic, such as 'NATO's Response to Hybrid Threats', which was organised by the NATO Defence College in April 2015 and was described as

NATO's 'largest-ever academic conference',[59] and 'Countering Hybrid Threats: Lessons Learned from Ukraine', the NATO Advance Research Workshop that took place in Bucharest on 28–9 September 2015.[60]

Three main developments can be identified in the second wave of the literature on Russian hybrid warfare in 2015. The first was an attempt to broaden the conceptual link between the previously established theories of hybrid threats and Russia's actions in Ukraine. The new discourse was characterised by a much deeper conceptual analysis of the connection between the theory of hybrid warfare and the Kremlin's actions in Ukraine.

While there were many different articles and papers that tried to extend the conceptual understanding of hybrid warfare in the context of the ongoing conflict in Ukraine, one of the most comprehensive conceptual works on the topic was carried out by András Rácz from the Finnish Institute of International Affairs.[61] After almost 100 pages of historical, theoretical and conceptual analysis of different aspects of asymmetric warfare, including an examination of different historical cases, such as Vietnam, Chechnya, Afghanistan, Iraq and Lebanon, as well as an in-depth analysis of Russia's actions in Ukraine, Rácz concludes that:

> All in all, concerning the terminology to be used when describing Russia's new mode of warfare as deployed in Crimea and in Eastern Ukraine, one may conclude that the expression hybrid war has become the most commonly used term. This occurred notwithstanding the fact that, strictly from the military science point of view, hybrid war originally had different, albeit related meanings. The fact that NATO adopted the term surely contributed to its wider use.[62]

Rácz was not the only one who questioned the relevance of the concept of hybrid warfare to the Russian intervention in Ukraine. For example, when discussing the application of the theory of hybrid warfare to Russia's actions in Ukraine, Freedman argues that:

> The Russian intervention has been described as an example of 'hybrid warfare'. This term gained currency after Israel was said to have been surprised and discomfited during the 2006 Lebanon War by the combination of guerrilla and conventional tactics adopted by Hizbullah … As with many similar concepts, such as asymmetric warfare, once adopted as a term of art it has tended towards a wider definition. Nor does it refer to a new phenomenon, for there are many examples in military history of the combination of regular and irregular forms of warfare.[63]

These works, based on in-depth theoretical and historical analysis, gave an early warning of the conceptual dissonance between the theory of hybrid warfare and its application to the Kremlin's actions in Crimea and eastern Ukraine. Yet the number of such articles paled into insignificance when compared with the amount of literature claiming that Russia has been waging a hybrid war in Ukraine.

The second main characteristic of this discourse was an attempt to explain Russian behaviour by interpreting Russian military publications and doctrines in general, and the well-known article published in 2013 by the Russian chief of the General Staff, General Valery Gerasimov,[64] in particular. By interpreting (or misinterpreting)[65] the so-called theory of new-generation war—as discussed by Gerasimov on several occasions,[66] as well as by two Russian officers, Sergey Chekinov and Sergey Bogdanov, in a series of articles published in *Voennaya Mysl'* (*Military Thought*), the major journal of the Russian General Staff[67]—many Western experts found a conceptual rationale for the Russian actions in Ukraine, arguing that the concept of new-generation warfare was how Russia conceptualised hybrid warfare.[68] As with the previous narrative, when some voices had warned that it would be misleading to label Russia's actions in Ukraine as hybrid warfare, this conceptual link between new-generation warfare and hybrid warfare also came under fire from several experts on Russian strategy. In November 2015, for example, Dima Adamsky argued that:

> When exploring the theory and practice of Russian operational art, terminology matters ... Applying the Western conceptual HW [hybrid warfare] framework to explain Russian operational art, without examining Russian references to this term, isolating it from Russian ideational context, and without contrasting it with what Russians think about themselves and others, may lead to misperceptions.[69]

Yet, despite this warning, the narrative that Russia was engaging in hybrid warfare in Ukraine—now supported by the idea that Russia had been developing this concept in the period before the Ukrainian Crisis—continued to occupy the centre of professional and academic debate.

The third major development in the second wave of this literature was a transformation of the 'call-for-action' narrative. In 2014, this narrative had been limited to rhetorical calls for help from the Eastern European members of NATO. However, from 2015 onwards, NATO members began to engage in a more constructive, practical debate about what the Alliance could or should do. This change of discourse was characterised by the realisation that NATO, as a military alliance, only had a limited capacity to respond to the multidimensional character of so-called Russian hybrid warfare. As one Estonian expert put it:

The Alliance will never embrace the full spectrum of challenges embodied in hybrid warfare and instead must remain focused on military issues ... [as] a strong allied conventional posture in the Baltics would be important not only in closing off easy conventional military opportunities for Russia, but also in countering the employment of hybrid tactics.[70]

Due to this realisation, in the period that followed, NATO implemented a number of military measures to enhance its capacity to respond to the 'the challenges posed by Russia'.[71] These included adapting the Readiness Action Plan (RAP) to increase NATO's military presence and activity in Eastern Europe and the Baltic States, and the creation of the Very High Readiness Joint Task Force (VJTF): 'a multinational brigade ... able to deploy within a few days in response to any threats or challenges that may arise on NATO's flanks'.[72] At the same time, NATO also recognised that military deployments would only offer a partial solution to hybrid threats and called for further cooperation between NATO and the EU and 'other relevant organisations that may play complementary roles in responding to hybrid threats'.[73]

2016—the third wave of discourse on Russian hybrid war

In a review of the literature produced on Russian hybrid warfare in 2014–15, one US expert, Michael Kofman, asked 'is 2016 the year we come to our senses? ... Today's conversation on Russia's use of hybrid warfare has become a discourse on something more arcane, resembling black magic. Generalisations about "Russian hybrid warfare" are not only unhelpful, but are becoming a cliché.'[74]

He was not the only one to doubt the applicability of the concept of hybrid warfare to Russia's actions in Crimea and eastern Ukraine.[75] Indeed, from 2016 onwards, numerous scholars added further criticisms to the doubts regarding the relevance of the concept that had already been espoused by Freedman, Adamsky, Rácz and others.[76] Bettina Renz and Hanna Smith, for example, argued that:

> 'Hybrid warfare' does not adequately reflect the content and direction of Russian military modernization ... [it] understates Russian ambitions and overestimates Russian capabilities at the same time. 'Hybrid warfare' oversimplifies Russian international politics/foreign policy, which is more complex than the label implies ... [and it] tells us nothing about Russian goals or intentions and mistakenly implies that Russian foreign policy is driven by a global 'grand strategy'.[77]

These doubts over the applicability of the concept to Russia's actions were accompanied by two interconnected narratives. The first was a call 'to think in Russian—in other words, to understand Moscow's motivations, and its understanding of the current confrontation'.[78] This attempt to understand Russian actions from the Russian perspective generated a vast amount of literature focusing on Russia's interpretation of the changing character of war. Though this analysis had already started in the previous year, as we saw above, 2016 would prove to be the year that analyses of Russian theoretical and conceptual military literature proliferated enormously.[79]

This focus on the Russian discourse on the changing nature of warfare was probably the main reason why scholars began to reject the applicability of the concept to Russia's actions, yet it also led to the second new narrative of the debate—a bifurcation of the analysis into two spheres: military and non-military. The discussion of Russia's military capabilities now turned back to more traditional forms of analysis emphasising the developments in Russian military 'hardware' (i.e. physical military capabilities),[80] and 'software'—the doctrinal and conceptual products of Russian military thought.[81] Simultaneously, the focus on Russia's use of non-military means and methods of confrontation led to a new conceptual focus in the West on the concept of information/cyber warfare.

The rise of Russian information/cyber warfare

Russian information/cyber warfare had been a topic of discussion for Western scholars since the late 1990s, with analyses of Russian military thinking and use of information and cyber operations in Chechnya and Georgia.[82] Though obviously important, this topic occupied a marginal place in the broader analysis of Russian military affairs until the beginning of the Ukrainian Crisis. However, since 2015 and the onset of the crisis, the subject of Russian information and cyber warfare has become a panacea for conceptual vagueness of hybrid war in relation to Russian actions not only in Crimea and eastern Ukraine, but also in the Baltic states, eastern Europe and elsewhere in the Western world.

The literature on Russian information and cyber warfare generally concentrates on the so-called 'Gerasimov Doctrine', the concept of 'new-generation warfare' and other Russian theories on the nature of modern conflicts, with the aim of trying to understand Russia's actions through the narrower lens of information and cyber warfare rather than hybrid warfare.[83] Compared with the efforts to explain Russia's actions via the concept of hybrid warfare, this attempt proved much more productive, as (unlike the concept of hybrid warfare) the theory of information warfare had been widely discussed in Russia, offering fertile ground for analysis of Russia's information/cyber warfare in Ukraine, the Baltic states and beyond. It also offered a conceptual link between the Soviet approach to conflict and contemporary Russian behaviour.[84]

Thus, following on from the tendency to reject the applicability of hybrid warfare to Russia's actions, attention has increasingly turned to Russia's use of information and cyber warfare. This focus has been led by the newly established NATO Strategic Communications Centre of Excellence in Riga, which published twenty-one research papers, articles and books, fourteen of which focus on Russian information and cyber activities, between January 2014 and the beginning of 2017.[85] However, the two leading voices in the debate have undoubtedly been Keir Giles and Mark Galeotti, both of whom have developed and promoted separate, if similar, narratives on Russian information and cyber warfare.

Keir Giles is an associate fellow of the Russia and Eurasia Programme at Chatham House and a director of the Conflict Studies Research Centre (CSRC), which focuses on Eurasian security. In 2015 and 2016, he published several articles and research papers in which he promoted the idea of Russian information and cyber warfare as the most useful conceptual lens through which to view and understand Russia's strategic behaviour.[86] Rooting his analysis in his extensive knowledge of the Russian theoretical, conceptual and doctrinal literature on information warfare and cyber operations, Giles highlighted the core differences between Russian and Western approaches to the issue, arguing that the West had largely failed to take effective actions to counter the Russian actions in this area.

Mark Galeotti is a senior researcher at the Institute of International Relations in Prague. Although he does not use the terminology of information or cyber warfare to describe Russia's actions, his arguments are otherwise very similar to Giles'. In his 2016 book, titled *Hybrid War or Gibridnaya Voina?*, Galeotti claimed:

> The West is at war. It is not a war of the old sort, fought with thunder of guns, but a new sort, fought with the rustle of money, the shrill mantras of propagandists, and the stealthy whispers of spies ... [this] political war that Moscow wages against the West, in the hope not of preparing the ground for an invasion, but rather of dividing, demoralising and distracting it enough that it cannot resist as the Kremlin asserts its claim to being a 'great power' and in the process a sphere of influence over most of the post-Soviet states of Eurasia.[87]

This is clearly similar to Giles' observations that:

'THE RUSSIANS ARE COMING'

Current trajectories indicate that further confrontation between Russia and the West is inevitable ... Crimea demonstrated that Russia does not have to wait until its military transformation is complete to use military force successfully. This is due to two key force multipliers: first, Russia's political will to resort to force when necessary ... and second, the successful integration of other strategic tools such as information warfare, reflecting the new doctrinal emphasis on influence rather than destruction.[88]

In their analyses, both authors establish a strong conceptual link between contemporary Russian strategic thinking and Soviet practices. Whereas Giles argues that:

> the techniques visible in and around Ukraine represent the culmination of an evolutionary process in Russian information warfare theory and practice, seeking to revive well-established Soviet techniques of subversion and destabilisation and update them for the internet age. For all their innovative use of social media, current Russian approaches have deep roots in long-standing Soviet practice.[89]

Similarly, Galeotti claims that:

> Russia's recent operations in Ukraine, especially the integrated use of militias, gangsters, information operations, intelligence, and special forces, have created a concern in the West about a 'new way of war' ... However, not only are many of the tactics used familiar from Western operations, they also have their roots in Soviet and pre-Soviet Russian practice.[90]

However, in making this conceptual link, it can be argued that Galeotti and Giles place too much emphasis on Russia's interpretation of contemporary conflicts as expressed in Russian/Soviet military and academic discourse. While the intention of both authors was to replicate the Russian approach to contemporary conflicts, it could be argued that they go too far in doing so, beginning to see the world via 'Russian glasses' rather than through a prism of critical analysis. For example, Galeotti's conclusion that 'the modern West—networked, globally-integrated, concerned with multiple real and perceived threats, and facing underlying crises of confidence and legitimacy—has specific vulnerabilities the Russians are eagerly exploiting'[91] bears more than a passing resemblance to Messner's concept of subversion-war, discussed in Chapter 3. Similarly, Giles' argument that 'an intellectual heritage permeated by postmodernist and relativist attitudes has now laid down fertile ground for disinformation and deception campaigns not only among Western historians, academics and even left-wing politicians, but among wide sectors of society',[92] clearly corresponds not only to Messner's concept of subversion-war but also to Panarin's theory of information warfare and Dugin's concept of net-centric war (Chapter 4). In other words, in their efforts to find a conceptual alternative to the concept of hybrid warfare in relation to Russian actions in Ukraine and elsewhere, Giles and Galeotti have ultimately brought the Russian interpretation of of confrontation between the West and Russia into the centre of Western academic and political debate, claiming that the Western 'responses to information warfare need to mirror some of the technical approaches adopted by Russia'.[93]

Until 2016, the discourse on Russian information and cyber operations principally focused on Russian actions in Europe, yet this began to change in 2016, as the US presidential elections of that year propelled the narrative into the centre of US political and academic discourse, reaching a peak after the leak of the Democratic National Committee (DNC) emails.[94] At this stage, given that Russia's attempts to undermine US democracy are still part of an ongoing political debate in the United States, it is impossible to draw firm conclusions on the subject at the time of writing (2017). However, it is possible to identify a central theme in the analysis of this issue.[95] In the words of a report issued by the US intelligence services:

> Russian efforts to influence the 2016 US presidential election represent the most recent expression of Moscow's longstanding desire to undermine the US-led liberal democratic order ... Russia's goals were to undermine public faith in the US democratic process ... [and its] campaign followed a Russian messaging strategy that blends covert intelligence operations—such as cyber activity—with overt efforts by Russian Government agencies, state-funded media, third-party intermediaries, and paid social media user or 'trolls'.[96]

This report provided very little evidence of the Kremlin's actions. Yet it is significant in the current context as it indicated that the ideas promoted by Giles, Galeotti and others have found their way into mainstream political discourse, not only in Europe but also in the United States. Hence the contemporary Russian interpretation of conflict as being comprised by non-military actions that undermine the political legitimacy of an adversary, rather than physically destroying it, has become a highly politicised element in the discourse on Russia's contemporary behaviour in the international arena.

The politicisation of Russia's hybrid/information/cyber warfare: analysing the debate

The politicisation of the Western discourse on Russia's approach to conflicts since the beginning of the Ukrainian Crisis has been led by three separate groups: (1) NATO in general and its Eastern European members in particular, (2) academia and (3) the Western political establishment.

As discussed previously, the Kremlin's reaction to the Ukrainian Crisis not only took the West by surprise but it also created a conceptual puzzle that Western military experts were keen to resolve. Thus the West was surprised not only by what the Russians did in Crimea and eastern Ukraine but also by how they did it, as their actions did not fit any of the Western concepts used to analyse contemporary conflicts. Since history teaches that people tend to fear an unknown that they struggle to understand or explain, especially in the context of the historical baggage in relations between the West and Russia, it is not surprising that the Western discourse surrounding this conceptual puzzle quickly became politicised, with different groups promoting different agendas.

As outlined above, NATO, and particularly its East European members, has been one of the main sources of the politicisation of Russia's hybrid/information/cyber warfare against the West. In his analysis of NATO's political discourse since the Cold War, Andreas Behnke argues that: 'The continued existence and relevance of NATO is to a significant extent contingent upon the Alliance's capacity to construct and maintain a cultural space called "the West" which provides its member-states with a common identity and purpose.'[97]

The dissolution of the Warsaw Pact in 1991 removed NATO's main adversary, which in turn led to a debate on the Alliance's purpose and identity. Though NATO's involvement in Afghanistan helped the Alliance to maintain its purposefulness, it did not solve the problem of its identity and its values. Since the so-called Global War on Terror was not about particular values or norms that NATO was established to defend, but about a universal sense of stability and resilience—and since some of NATO's partners in this war are not part of the West and do not necessarily share its values—the Alliance's participation in the war against terror simply increased its identity crisis rather than solving it.[98] Moreover, the economic crisis that struck Western economies at the end of the 2000s further enhanced the sense of insecurity among NATO's leadership, as a 2012 research report prepared by the RAND Corporation stated: 'Financial and economic constraints are redefining NATO's ability to provide security in the coming decade. NATO faces more than a simple, short-term budget squeeze: It is confronted with a secular trend that will have a serious impact on NATO Europe's ability to deploy and sustain power over long distances.'[99]

In the context of this identity crisis and general economic uncertainty, the revival of the Russian threat has proven convenient to NATO's leadership at the rhetorical level. Conceptualising and politicising Russian actions in Crimea and eastern Ukraine as a new hybrid threat that endangers the West by compromising its core democratic values has served NATO's interests in three ways. The first is the revival of the concept of hybrid warfare itself, forcing NATO's member-states to commit resources to a whole spectrum of activities that rest outside traditional military challenges. One of the results of this revival, for example, was the establishment of the NATO Strategic Communication Centre of Excellence in Riga, the creation of which had long been discussed, but on which NATO had dragged its feet since 2013.[100] As the 'Wales Summit Declaration' states: 'We will ensure that NATO is able to effectively address the specific challenges posed by hybrid warfare threats [and therefore] we welcome the establishment of the NATO-accredited Strategic Communications Centre of Excellence in Latvia as a meaningful contribution to NATO's efforts in this area.'[101]

The second way in which the idea of Russian hybrid/information/cyber warfare has served NATO's interests is that it was used to strengthen the institution's identity as the main defender of Western values. The narrative of Russia's revival as a threat to these values has consequently been used as a means to resolve NATO's post-Soviet identity crisis. As Michael Kofman put it:

> Individually, Western countries are knowledgeable about the extent of Russian political influence in their respective nations; but collectively the West has chosen to speak in narratives, and paint a caricature of how Moscow uses its instruments of national power. That is understandable as a part of an effort to motivate NATO, raise alliance awareness, and reassure vulnerable members.[102]

The third factor behind the politicisation of the discourse on Russia's actions is NATO's East European members, which have been the most vocal voices in this process of redefining Russian actions in Crimea and eastern Ukraine as a new threat to the security and integrity of the Western world. Due to a genuine historical fear of their powerful eastern neighbour, political and military experts from these countries have sought to gain additional protection from their more powerful allies in the West. However, it could also be argued that the efforts of these states to politicise Russia's hybrid/information/cyber warfare as the main threat to Western values have also been about gaining access to financial and military support, thus improving not only their traditional military power but also enforcing their political legitimacy, especially when it comes to the Baltic states. For example, Estonia and Latvia, both of which have large Russian-speaking minorities, have long been concerned with the threat posed by the Kremlin's propaganda, traditionally describing the potential Russian subversion of their own current political establishments as a 'coup attack'.[103]

In other words, following the view of Lithuanian president, Dalia Grybauskaitė, NATO is an 'insurance policy' that 'does not mean you will get sick, but it is better to have it', and therefore it should protect from any irredentist actions Moscow may implement, directly or indirectly.[104] While the ability of NATO to protect its Baltic members in the event of Russian conventional offence is open to question,[105] the narrative that NATO's eastern flank is vulnerable to overt or covert Russian actions has been used to serve the purposes of both the Alliance and its eastern members. On the one hand, NATO used this narrative to establish the VJTF as a response to the new challenges 'that arise, particularly at the periphery of NATO's territory', thus enhancing its members' political and economic commitment 'to act together and decisively to defend freedom and our shared values of individual liberty, human rights, democracy, and the rule of law'.[106] On the other, from the perspective of the Baltic states, the permanent presence of NATO forces on their territory offers evidence of the promised 'insurance', thereby enforcing the current political establishments of these states in the eyes of their own populations. Whether VJTF is able to prevent, or even deter, Russian's non-military activities in these states, such as information and cyber operations, is a completely different story.

Academia is the second source of the politicisation of the narrative of Russian hybrid/information/cyber warfare against the West. However, before analysing this politicisation process, it is important to emphasise that an integral part of the debate was a genuine attempt to understand the complex combination of military and non-military means and methods employed by Russia in Crimea and eastern Ukraine.

Describing the relationship between academics and policy-makers, Freedman claims that 'when a new policy problem emerges ... there will be a surge of activity as grants become available'.[107] As discussed previously, Russia's actions in Crimea and eastern Ukraine surprised and puzzled Western experts, and since this topic has been in vogue since 2014, many academics have produced analysis, commentary and research on the topic. Regardless, the type of research conducted by these academics, whether it was of pure conceptual nature,[108] of with more political emphasis, trying to direct policy-making by promoting,[109] (or challenging)[110] the alarmist idea of Russian hybrid warfare as a threat to the Western world, they all contributed to the politicisation of the topic, keeping it at the centre of Western discourse.

Academics rarely agree with each other, thus creating a challenging and productive interaction between academia and policy-making. Yet, conversely, this interaction between academics and policy-makers, as seen in the case of Russian hybrid warfare, seems to have created an unhelpful situation where, in Freedman's words, 'challenging interaction [turns] into an echo chamber' in which officials incline towards academics who share their assumptions and reinforce their conclusions. 'It is not that the encounter with power turns the academic away from truth,' Freedman writes, 'but those with a particular interpretation of the truth are more likely to get a hearing.'[111]

The third major source of the politicisation of Russia's hybrid/information/cyber warfare against the West has been the Western political establishment. As with any narrative within the political domain, the idea of Russia's intervention in domestic affairs has been used by different political groups for political purposes. For example, in the UK the Kremlin was blamed for influencing the Brexit referendum,[112] in Germany Russia stood accused of subverting the re-election chances of Chancellor Merkel,[113] and, in the United States, Russia has been seen as having undermined the Democratic Party candidate Hillary Clinton in order to secure the election of Donald Trump.[114]

The Western political establishment has generally embraced the alarmist agenda promoted by NATO and certain academic experts. The best examples of this were expressed in a 2016 report produced by the European Union Institute for Security Studies (EUISS) for the European Parliament and the resolution that followed the report. Titled 'EU Strategic Communications with a View to Countering Propaganda', the report states that: 'Over the past few years, the EU has been increasingly hit by destabilising messages amounting ... to coherent hostile "strategic communications" campaigns. Those promoted and orchestrated by Russia ... have explicitly targeted the EU as such, its nature and its policies.'[115]

Based on this report, the European Parliament approved a resolution claiming that 'Russian strategic communication is part of a larger subversive campaign to weaken EU cooperation and the sovereignty, political independence and territorial integrity of the Union and its Member States.'

The resolution further criticised: 'Russian efforts to disrupt the EU integration process and deplores, in this respect, Russian backing of anti-EU forces in the EU with regard, in particular, to extreme-right parties, populist forces and movements that deny the basic values of liberal democracies.'[116]

'THE RUSSIANS ARE COMING'

Hence the Western political discourse around the so-called Russian hybrid threat echoed the narratives and the language promoted by NATO and some members of the academic community. And as with any political narrative, politicians have used this renewed alarm over Russia's actions to promote their own agendas, the best example of this being the alleged link between Trump's presidential campaign and the Kremlin, which has been used both by Trump's supporters and his opponents. Thus, according to the report by the US intelligence agencies, cited earlier:

> Russian President Vladimir Putin ordered an influence campaign in 2016 aimed at the US presidential election, the consistent goals of which were to undermine public faith in the US democratic process, denigrate Secretary Clinton, and harm her electability and potential presidency. We further assess Putin and the Russian Government developed a clear preference for President-elect Trump. When it appeared to Moscow that Secretary Clinton was likely to win the election, the Russian influence campaign then focused on undermining her expected presidency.[117]

Yet only a month later, White House Chief of Staff Reince Priebus claimed that: 'The top levels of the intelligence community have assured me that [the allegation] is not only grossly overstated, but also wrong [as] they have made it very clear that the story is complete garbage.'[118]

As this debate is ongoing at the time of writing in 2017, it is difficult to draw any firm conclusions on its causes or true aims; all that can be said with any certainty is that different political actors in the West are using this narrative of the Russian threat to promote their agenda and delegitimise that of their political adversaries. Ultimately, therefore, the initial narrative surrounding Russia's actions in Crimea and eastern Ukraine has been politicised for domestic purposes, creating an image of Russia as the principal threat to Western democratic values and using this alarmist image for domestic political gains.

Conclusions: 'the Russians are coming'

As has been shown throughout this chapter, the Western discourse on Russian hybrid warfare, together with its information and cyber elements, contains two main aspects: the conceptual and the political. At the conceptual level, as Russia's actions in Crimea and eastern Ukraine surprised Western experts, the debate over how best to explain them developed over two successive stages. The initial attempt involved applying the concept of hybrid warfare to Russian behaviour and tactics. Yet, in doing so, many observers came to the conclusion that the concept provided a misleading image of Russia's actions. By 2016, this effort to understand the Russian approach to contemporary conflicts through a Western conceptual framework had generally been abandoned, as an increasing number of researchers and experts began to accept Adamsky's conclusion that 'utilizing Western terms and concepts to define the Russian approach to warfare may result in inaccurate analysis of Russian modus operandi'.[119] Or, as Timothy Thomas stated: 'Simply overlapping Western concepts on Russian thinking doesn't always work. If evolving foreign concepts are not understood from their local context, then the West will always be chasing after outliers without understanding where they fit in Russia's overall theoretical and planning process.'[120]

This in turn led to a different focus, in which a Western conceptual framework was replaced by a conceptual framework that sought to understand the nature of contemporary conflicts through the prism of Russian military thought, thus unintentionally (or intentionally) promoting Russia's interpretation of regional geopolitics.

In tandem with this conceptual debate, a political discourse has been taking shape. NATO in general, and its Eastern European members in particular, have used the narrative surrounding Russia's hybrid/information/cyber warfare against the West to close ranks against a common new/old enemy after a twenty-five-year search for self-identification accompanied by shrinking budgetary resources. The resurrection of the Kremlin as the great evil, seeking to subvert Western democratic values, has offered NATO members not only political unity amongst its members and their financial support, but has also revived its historical purpose and identity by pressing the pressure points of Cold War nostalgia for a clearly defined evil that uses dirty tricks to achieve its villainous aims.

In the initial Western political discourse, Russia's actions in Crimea and eastern Ukraine were framed as a problem for international relations, international law and international security. As President Obama put it in 2015:

> We recognise the deep and complex history between Russia and Ukraine. But we cannot stand by when the sovereignty and territorial integrity of a nation is flagrantly violated. If that happens without consequence in Ukraine, it could happen to any nation gathered here today. That's the basis of the sanctions that the United States and our partners impose on Russia. It's not a desire to return to a Cold War.[121]

However, despite Obama's reassurances, the narrative of Russia's subversion of Western democracy has been politicised by different political groups for their own domestic purposes. In March 2016, commenting on the hysteria among America's European allies with regard to Russia's apparent use of hybrid warfare, Kofman argued that the sense of alarm in those countries was almost akin to 'a modern version of America's red scare from the 1940s and 50s'.[122] One presidential election later, it seems that the US has been going down the same path. The ongoing political debate about Russia's threat evokes the notorious phrase attributed to US Secretary of Defense James Forrestal: 'The Russians are coming'.

As this chapter has shown, the debate over Russia's use of military and non-military means and methods in Crimea and eastern Ukraine has evolved into a discourse that has been used for instrumental purposes in domestic politics; from being a problem of international order, in the context of a geopolitical struggle, the discourse has consequently broadened into allegations that Russia is attempting 'to weaken and split the EU'[123] and 'to undermine the US-led liberal democratic order'.[124]

7

'*GIBRIDNAYA VOYNA* AGAINST US IS COMING'

THE POLITICISATION OF WESTERN *GIBRIDNAYA VOYNA*

As we saw in Chapter 5, the concept of *gibridnaya voyna* is informed by ideas that have been at the centre of academic and military discourse in Russia since the mid-1990s. While Part 2 of the book focused on the conceptual development of the idea of the 'Western *gibridnaya voyna* against Russia', this chapter discusses how the different narratives represented by this idea influenced Russian military thought and were politicised by the Russian political establishment. The first part of the chapter examines the development of the theory of new-generation war, which many Western experts claim shares similar characteristics to the concepts of subversion, net-centric and information warfare (and, therefore, the concept of *gibridnaya voyna*). The second part of the chapter focuses on the politicisation of these ideas, and on the ways in which the Russian military and the Russian political establishment have used the narrative of *gibridnaya voyna* to achieve their political aims.

The rise of new-generation warfare—Reading Sergey Chekinov and Sergey Bogdanov

The theory of new-generation warfare is closely associated with the work of two Russian officers, Colonel (ret.) Sergey Chekinov and Lieutenant General (ret.) Sergey Bogdanov, both from the highly influential Centre for Military and Strategic Studies of the General Staff of the Russian Federation Armed

Forces (Chekinov is the centre's director whereas Bogdanov is its lead researcher).[1] Both authors have published hundreds of articles and papers (separately or together) on different subjects related to military affairs. Since the late 2000s, their joint publications on the changing nature of contemporary conflicts have played a vital role in shaping the views of the Russian military establishment in general, and particularly the views of the Russian chief of the General Staff, General Valery Gerasimov.[2] However, before discussing Chekinov and Bogdanov's concept and its relationship to the idea of *gibridnaya voyna*, it is important to analyse their views on the phenomenon of war more generally in order to shed some light on the broader theoretical framework that has influenced Russian military thought since the beginning of the twenty-first century. As Chekinov and Bogdanov started to write about new-generation warfare long after the ideas of Messner, Dugin, Panarin and their proponents had begun to influence Russian academic/military discourse, they tend to react to these ideas rather than trying to shape them. In other words, in their discussion of the concept of *gibridnaya voyna*, the two authors have tried to reflect on the nature of the concept while integrating existing ideas into their vision and understanding of warfare, rather than influencing the characteristics of the concept itself.

The first and most important aspect of Chekinov and Bogdanov's work is their discussion of the continuous transformation of war as a social-political and military phenomenon. An analysis of this discussion offers a highly interesting insight into the way that they understand this phenomenon and how they have developed this understanding in their publications. For example, in one of his earlier articles, referring to the Clausewitzian definition of war as a continuation of politics, Chekinov argues that:

> Since Clausewitz ... mankind moved forward two centuries. The wars of the 19th and 21st centuries will significantly differ one from another not only in content but also in essence. Contemporary war is a qualitatively new phenomenon, the analysis of which requires a re-evaluation of the existing concepts of social development and a refinement of the conceptual structure.[3]

However, when referring to the same topic in 2017, Chekinov and Boganov argued that: 'In our view, it is too early to change the interpretation of the concept of "war." It has always been and will necessarily be waged with the use of certain types of weapons and constitute an organised armed struggle of peoples, states, alliances, [and] coalitions.'[4]

This apparent inconsistency is not a contradiction but a development of their conceptual understanding of the role of non-military means and methods in contemporary conflicts, as well as the conceptual difference between 'war' and 'political confrontation'. As they put it:

> The criterion for the presence or absence of war is the place and role of military and non-military means in a political confrontation. If an armed struggle or other actions of the armed forces are the main means for achieving political goals, and all other non-military means of violence are directed to maximise the effect of employed armed forces, it is ... a war. Following this, a political confrontation, in which the emphasis is on non-military means of violence, and the troops only have an impact because of their mere presence or an action without the use of fire and strikes, but by demonstration [of force], threats [to use it] and others—it is not a war.[5]

Since the main emphasis of *gibridnaya voyna* is on non-military means and methods, it is important to trace their role and place in Chekinov and Bogdanov's work. By analysing the articles they have produced, whether as co-authors or independently, it is possible to point to two main conclusions that shaped their conceptualisation of the role of non-military means in contemporary conflicts. The first is related to 'war' and is based on their analysis of US-led operations since the end of the Cold War. For example, writing about the US-led operations in Iraq in 1991 and in Yugoslavia in 1999, Chekinov argued that:

> The terms of victory in war are changing. The principle of achieving the military–political goals of wars and armed conflicts with minimum human and material losses comes to dominate the military theory and practice of advanced ... states. Following this principle, military actions are directed not to capture territory and destroy enemy personnel, but, above all, to undermine its will and ability to resist, [thus] enforcing peace under the conditions imposed on [the enemy].[6]

Further developing this idea, Chekinov argues that his analysis of these conflicts reveals that the erosion of the enemy's will was achieved by a combination of non-military means led by information and psychological operations: 'The aim of information–psychological operations is to substantially weaken the enemy's military capabilities by non-violent methods that target its information processes, thus misleading and demoralising its population and members of its armed forces.'[7]

As such, according to Chekinov and Bogdanov, 'war' includes non-military actions based on the kinds of non-violent methods that would usually precede the beginning of military activities:

> Based on the results of the analysis of contemporary conflicts and local wars that have been led by the U.S. and its allies during the past decades, it is possible to

assume ... that a country, which plans an aggression, ... will seek to disorganise the systems of civil, military and operational control of the country, against which the aggression is planned, even during peace-time or in the initial stages of aggression ... In this same period, the governments and the military leaderships of confronting states will employ a whole range of political, economic, military–organisational actions, as well as dealing with moral–psychological and information-disguising problems.[8]

This understanding, however, is not unique to Chekinov and Bogdanov, as it is widely shared by many other Russian strategists and military thinkers. For example, Andrey Kokoshin argues that:

Information–psychological influence [and] special propaganda are intended first and foremost to serve the task of suppressing the enemy's will to resist [and] the disruption of control at all major links, but by other means—without direct physical influence on the enemy's personnel ... In the contemporary environment of the different types of limited wars and armed conflicts, peacemaking and peace-enforcing operations; information–psychological efforts, in general, and special propaganda, in particular, have begun to play an even more important role.[9]

Similarly, Nikolay Volkovsky, who has analysed the history of information wars since antiquity, concludes that:

Today, more and more efforts and means are involved in information warfare, significantly increasing its consequences. The wars and armed conflicts of the last decades of the 20th century and the beginning of the 21st century serve as a clear illustration of this. Their analysis shows without any doubt that in the contemporary world the course and outcome of military actions of all sizes is determined by the art of information warfare.[10]

Chekinov and Bogdanov's second conclusion regarding the non-military means and methods that have been used in contemporary conflicts is related to what they define as 'political confrontation'. Based on an analysis of the geopolitical transformations that have taken place since the Cold War, as well as associated processes such as globalisation, economic and financial interdependence, and the technological revolution in communications,[11] they argue that:

Under the conditions of the globalisation of world processes, the enormous economic superiority of leading powers and the heavy financial dependence of the majority of the other countries on them [the leading powers], there is no objective need to conduct large-scale wars. Such wars are not expected because of the threat of catastrophic consequences of the use of nuclear weapons, on the one hand, and on the other—because new ways and means have been found of achieving political and strategic objectives by conducting local wars [and] conflicts; by political, economic [and] informational pressure; and by subversive actions inside the adversary state.[12]

Hence, according to Chekinov and Bogdanov, though non-military means and methods can feature in both of these two scenarios—'war' and 'political confrontation'—there is a clear distinction between them. In the former, they are intended to soften the enemy before large-scale military operations, whereas, in the latter, they are intended to be a substitute for military deployment. It was this first scenario that the pair conceptualised as new-generation warfare and the second as *gibridnaya voyna*. Although this chapter focuses on the latter, it is also important to briefly outline the main features of new-generation warfare, especially as Western academic experts tend to confuse the two, misinterpreting the distinction defined by Chekinov and Bogdanov.[13]

In 2013, the pair published one of their most widely read articles, entitled 'The Nature and the Content of the New-Generation War', in which they introduced their concept of new-generation war.[14] On the basis of this article, many Western experts came to the erroneous conclusion that the concept of new-generation warfare was the same as the concept of hybrid warfare, or *gibridnaya voyna*.[15] In fact, as noted above, new-generation war and *gibridnaya voyna* are two separate things, as the first is war and the second is not.

The authors emphasised that non-military methods and activities, and especially information–psychological operations, are integral parts of new-generation warfare:

> In a new-generation war, a leading role will be taken by the information–psychological struggle, directed to achieve superiority in the sphere of command and control, as well as to supress the morale of the military personnel and the population of the adversary. In the [contemporary] environment of the information–technological revolution, this information–psychological struggle will create [the required] preconditions for achieving victory.[16]

Accordingly, these actions are intended to soften the enemy as a preparatory stage before a war, as 'new-generation war [is] an international armed conflict' planned ahead of time by the offensive side and, therefore, before it begins, in an attempt 'to create favourable military–political and economic conditions for the employment of the armed forces'. Such an offensive would require a joint plan of large-scale actions to be implemented, including 'all types of different confrontations: informational, morale–psychological, ideological, diplomatic, economic and others'.[17]

To clarify the conceptual difference between new-generation war and *gibridnaya voyna*, Chekinov and Bogdanov argue that *gibridnaya voyna* is 'an element of interstate confrontation intended to realise the national interests of the state by an extensive use of indirect actions, while maintaining the armed forces as a deterrent':

> The experience of wars and armed conflicts during the last decade shows that the intensity of the confrontation in spheres, other than the sphere of armed struggle, has significantly increased. This often led to the achievement of the intended aims by non-violent means without the use of military force, i.e., by *gibridnaya voyna*.[18]

This distinction between an actual war and *gibridnaya voyna*, which has also been described as war (*voyna* meaning war in Russian) in contemporary Russian military and academic discourse, was best formulated by another prominent Russian strategist, General Makhmut Gareev. Gareev criticised the increasing fashion for talking about subversion/net-centric/information/hybrid wars in Russian academic and military debate:

> In recent years, due to an increase in international confrontations of the proportion of political, diplomatic, economic, informational, cyber, [and] psychological means and methods of achieving political objectives, raises the question of a radical change in the concept of war ... It has always been considered ... that a confrontation in any field without arms—is a struggle; but a continuation of politics by violent means, with the use of armed violence—is a war. Only problem is that some of our philosophers, who do not know history well, believe that all of these non-military means have appeared only today and, on this basis, assume that the use of such means is a war.[19]

While Russian military thought does not consider *gibridnaya voyna* a form of war, despite its misleading name, Chekinov and Bogdanov accord it a very special role in the formulation of strategy. In their discussion of the role and place of strategic deterrence, for example, the two authors argue that strategic deterrence should be planned and executed by both the civil and military leadership of a given country, based on a combination of 'military (violent) and non-military (non-violent) actions'. Whereas the former should be the responsibility of the armed forces, the latter should be 'implemented by federal executive bodies'.[20] As a result, *gibridnaya voyna* is not the direct responsibility of the armed forces, though they might be used as a tool that supports political–diplomatic, economic, information and other non-military actions simply by their presence or by the demonstration of military potential.[21] Instead, *gibridnaya voyna* is the responsibility of the civilian leadership, as its main purpose is 'not to defeat the enemy on the battlefield, but to create in its consciousness "an image of defeat"',[22] by a combination of non-military means defined as: 'A set of public, [and] social institutions (organisations); political, legal and economic norms; cultural values; [and] public information and technical systems—[all] used by the state to influence the internal and external relations between states.'[23]

This conceptualisation of non-military means in general, and *gibridnaya voyna* in particular, as formulated by Chekinov and Bogdanov, clearly resembles many of the ideas proposed by Dugin, Panarin and the proponents of Messner's subversion-war.[24] Although Chekinov and Bogdanov reject the idea of calling these activities 'war', they praise them as an integral part of the contemporary geopolitical environment with indispensable and ever-increasing importance: 'today, during confrontations between states, as well as between rival forces within them, the centre of gravity has been visibly shifting towards non-military means due to the increasing danger of mutual destruction'.[25]

Moreover, Chekinov and Bogdanov clearly recapitulate the shared narratives of Dugin's net-centric war, Panarin's information war and Messner's subversion-war when discussing the history of the Western use of non-military means and methods against Russia and the required countermeasures. Although they consistently emphasise that their analysis is based on the performance of the US military and its allies since the early 1990s,[26] their conceptualisation of non-military means, as a distinctive and non-violent way of political confrontation (i.e. *gibridnaya voyna*), is based on an analysis of the Cold War and the defeat of the Soviet Union as a result of Western non-military actions:

> The complex application of non-military means by the U.S. in a 'cold war' against the Soviet Union yielded more significant results than the use of military force ... Today, it seems obvious that at the end of the twentieth century American civilisation has secured the dominant position on the entire planet by using the methods of information–psychological influence on people and sociopolitical bodies of neighbouring countries ... At the end of the 20th century, methods of information–psychological influence have allowed the Americans to destroy the Soviet Union.[27]

As with the narratives promoted in Panarin's concept of information warfare and Dugin's theory of net-centric warfare, Chekinov and Bogdanov argue that the West's non-military offensive against Russia did not come to an end with the Cold War; instead, 'it continues and moves forward onto a new stage by the employment of the "colour revolution" elements'. Thus:

The means of the Cold War have been supplemented by the elements of traditional war, as well as by new forms of influence, such as subversive operations on an enormous scale in the spheres of politics and economics. Elements of the 'colour revolution' have spread to other regions of the world—the Middle East, the Balkans. There is resolute evidence to believe that humanity has entered a new era of global war, a new type of war.[28]

Following this argument, the two authors accurately echo the conceptual foundations of the Russian way of thinking about *gibridnaya voyna*, as promoted by Dugin, Panarin and the advocates of Messner's subversion-war: 'The effectiveness of this new type of war ("peaceful" war) has convincingly been proven in splitting the population of the adversary state into rival parts, in creating an obedient "fifth column", in inclining one part (rebellious) to a betrayal and the [following] capture of the power by pro-Western activists.'[29]

According to Chekinov and Bogdanov, the West still considers Russia its main threat and adversary, and the Western leadership 'will calm down only when our country and our people will be reduced to a state that deserves [nothing but] mockery and contempt'.[30] Therefore, it is not surprising that both authors, much like Panarin and Dugin, suggest that Russia should develop its own arsenal of non-military means and methods. It is possible to point to two main explanations in their work for the usefulness of *gibridnaya voyna* methods for Russia's strategy.

The first is the contemporary geopolitical environment. Chekinov and Bogdanov argue that 'a direct military threat to the Russian Federation from the U.S. and NATO member-states in the near-future is unlikely'.[31] Yet in their analysis of the different trends that have been occurring since the end of the Cold War—such as globalisation, the increasing interdependence of national financial and economic systems, the information revolution—Chekinov and Bogdanov argue that such trends have caused 'fundamental changes in the employment of political, economic, [and] indirect actions, as well as in the usage of non-military means, for resolving contemporary interstate contradictions'.[32] Based on this conclusion, the authors offer a clear policy recommendation, claiming that 'taking into consideration the contemporary realities, it will be advisable for Russia to define, and then implement, the strategy of indirect actions, for the lack of other options, as the state strategy'.[33]

Although this recommendation regarding the use of indirect actions could apply to any country, their second recommendation is specific to Russia, which they claim should seek to adopt non-military, indirect actions in its attempt to counteract the 'American ambition to impose its will, approaches, values, [and] the unipolar system of world-order upon the members of the international community'.³⁴ After analysing the development and prevalence of asymmetric conflicts in the second part of the twentieth century, as well as the different conceptual and theoretical frameworks used to explain these conflicts, Chekinov and Bogdanov come to the seemingly self-evident conclusion that 'an employment of asymmetric actions frequently allows a weak adversary to achieve political victory'.³⁵

Taking their analysis and examination further, by basing their conceptualisation on the works of numerous strategists and military thinkers, from Carl von Clausewitz, Liddell Hart and Mao Tse-tung to Henry Kissinger, the pair conclude that 'from the perspective of [military] science, it becomes obvious that the terms "asymmetric approach" and "asymmetric actions" are very close in their content to the terms "indirect actions" and "the strategy of indirect actions"'.³⁶ Combining this conceptualisation with their understanding that Russia is the weaker side in its struggle against the more powerful West, and particularly the United States, they go on to argue that: 'To deter and prevent aggression from any state (coalition of states) and to ensure the military security of the Russian Federation, it is advisable to take asymmetrical measures, which should be systematic, comprehensive, [and] combine political, diplomatic, information, economic, military and other efforts.'³⁷

In other words, Russia should resort to indirect non-military means and methods as part of its strategy. They give three main reasons for doing so. The first is the fact that the geopolitical environment at the end of the twentieth and beginning of the twenty-first century has been developing in a way that makes non-military actions a more preferable tool in inter-state confrontations, especially when the confrontation is between nuclear powers fearful of the danger of escalation once direct military actions have been used. The second is their belief that the West did not stop its non-military, indirect offensive against Russia in the post-Cold War period. Finally, the third reason is that, since the struggle between Russia and the West is asymmetric (where Russia is the weaker actor), these indirect (asymmetric) activities will serve to compensate for Russia's weaknesses.

Chekinov and Bogdanov have made a number of important contributions to the Russian military discourse surrounding the theory of *gibridnaya voyna* and its claim that the West is engaged in a non-military subversive offensive against Russia. The first is the clear conceptual distinction they draw between new-generation war and *gibridnaya voyna*: whereas non-military (i.e. economic, diplomatic, financial, information, cyber) means and methods are employed in both cases, in new-generation warfare these methods are intended to prepare the ground for subsequent military actions, while in *gibridnaya voyna* they are used for a stand-alone, non-violent political confrontation. Hence a recurring theme throughout their work is a critique of the use of the term 'war' for actions and methods that are not intended to precede or accompany a direct deployment of armed forces (*gibridnaya voyna*): 'The use of armed forces is a prerequisite condition for determining a stage of an interstate confrontation as a war. Otherwise, it can be defined as anything else—"political conflict", "incident", "competition"—but not as a war.'[38]

By drawing this conceptual distinction between new-generation war (to which the term 'war' applies, properly speaking) and *gibridnaya voyna* (which is not an actual war), Chekinov and Bogdanov accept and promote *gibridnaya voyna* as a legitimate phenomenon in inter-state confrontations.

The second contribution is their reference to political theories and narratives that have been at the core of Russian academic and military discourse for more than a decade to analyse and discuss the role and place of *gibridnaya voyna* as a combination of non-military indirect means and methods intended to subvert and weaken political adversaries without resorting to military actions. Though they do not refer directly to the ideas of Panarin, Dugin and Messner, it is safe to assume that they are more than familiar with their works. As discussed above, the narratives of information, net-centric and subversion wars can be traced throughout Chekinov and Bogdanov's publications. However, while they have criticised the conceptualisation of these narratives as a war, they agree with the existence of the phenomenon. In other words, Chekinov and Bogdanov, who ultimately represent the general course of Russian contemporary military thought, have promoted and institutionalised the idea that the West, and especially the United States, has been trying to subvert and undermine Russia by employing different non-military means and methods (i.e. *gibridnaya voyna*) and that Russia, therefore, should seek to do the same.

The politicisation of gibridnaya voyna and its narratives in Russian academic and military discourse

Many Western scholars view Chekinov and Bogdanov as the fathers of the conceptualisation of non-military indirect means and methods in Russian contemporary military thinking.[39] Yet an analysis of Russian academic and military discourse shows that this is not the case. For example, as early as 2000,

'GIBRIDNAYA VOYNA AGAINST US IS COMING'

Vladimir Serebryanikov and Alexander Kapko, a retired lieutenant general and a former high-level diplomat, argued that:

> At the end of the 20th century, the non-military means have equalled the military ones in terms of their effectiveness, if not even exceeded them. It has become possible to achieve by non-military means certain goals that could be previously achieved only by a bloody war. Without direct use of military force, but as a result of the integrated use of political, diplomatic, economic, informational, ideological–psychological, intelligence, covert and other non-military means, the West destroyed the Soviet Union, the Warsaw Pact, Yugoslavia and Czechoslovakia, including their security systems, law enforcement structures, [and] armed forces.[40]

Consequently, in practice, the main contribution of Chekinov and Bogdanov was the previously discussed conceptual division between the theory of new-generation warfare and the idea of *gibridnaya voyna*, rather than the introduction of the idea of non-military means and methods to Russian academic and military discourse per se. Put differently, the idea that non-military means and methods such as 'economic, information, ideological, intelligence, [and] covert subversive activity against a state' have the potential to be more powerful and achieve better results—especially in the era of mass media, capable of 'reaching every person on the planet conquering his soul and heart [and] turning him into a supporter', when globalisation has weakened 'the control of states over their economic, information [and] spiritual processes', and the 'increasing danger of self-destruction in large-scale military confrontations has compelled to increase the stakes on non-military means'[41]—these concerns have been a part of Russian academic and military discourse for more a decade before Chekinov and Bogdanov conceptualised their theory of new-generation warfare.

In analysing the military theoretical discourse, since the early 2000s, on the role and place of non-military means and method in interstate confrontations, one can easily trace several very similar narratives between the idea of *gibridnaya voyna* (as discussed by Chekinov and Bogdanov) and Panarin's information war, Dugin's net-centric war and Messner's subversion-war, all of which had been at the centre of academic debate since the late 1990s. The first is that, since the late twentieth century, the evolving geopolitical situation, combined with the information and communications technological revolution, has increased the importance of non-military means, thus replacing direct military confrontations with economic, diplomatic, information, cyber and other indirect methods. The second theme is that these techniques had already been mastered by the West, and especially the United States, and were used against the Soviet Union, thus leading to the latter's defeat in the Cold War.

Analysing the ways in which the US and NATO have been conducting their conflicts since the end of the Cold War, as well as their doctrinal theoretical publications, Russian military and academic scholars have developed a third major narrative—the idea that the West, as a part of its vision of a world-order dominated by Western civilisation, still implements a complex of non-military means and methods intended to compromise and subvert the Russian political establishment. This in turn leads directly to the fourth theme or narrative in Russian military thought, and probably the most important of this discourse—Russia's political–military leadership needs to invest resources and efforts not only to defend Russia against the West's use of subversive methods but also to develop a systematic approach to the question of 'how and in what way non-military means, especially socio-political and economic, will eliminate the military threat to Russia.'[42]

Regardless of the given title, whether information warfare, net-centric warfare, subversion-warfare or *gibridnaya voyna*, or simply the 'implementation of non-military indirect means and methods', these concepts share and promote the same narratives, thus creating a monolithic discourse that has evolved since the late 1990s, a full analysis of which is beyond the scope of this book. However, it is nevertheless important to trace the general development of this discourse, and in particular, to focus on the main sources of its politicisation.

When analysing the discussion surrounding *gibridnaya voyna* and its associated narratives in Russian discourse, it is difficult to avoid noticing that it has become an increasing source of debate in the twenty-first century. In 2000, commenting on the level of attention that non-military means and methods had received in Russia, Serebryanikov and Kapko stated:

> Today, even in the highest echelons of the state leadership, it is difficult to find an expert who is professionally engaged with the problems of the use of non-military means ... In the past 10–15 years, only a few popular articles appeared in this field. Neither aims, nor subjects of the study are defined. Domestic and foreign experiences are not summarised. There are no scientists, who specialise in this field.[43]

As of 2017, this situation has changed dramatically. As has been discussed throughout this chapter, as well as in Part 2, a vast amount of research has been conducted, various articles and books have been published, and research centres and knowledge hubs have been established. Different theoretical and conceptual approaches have been developed, criticised and reconceptualised, some becoming the latest conceptual fashion while others have faded into obscurity. Most of these ideas have already been discussed in this book, though there are also others that have received far less attention, such as the theory of 'controlled chaos'[44] and the concept of 'colour revolution'.[45] However, despite the variations in how these concepts and theories are labelled, they have all promoted and politicised a common narrative: Russia is facing a Western offensive that is intended to destroy Russia by political subversion via information–psychological, ideological, economic and other non-military dimensions, as it had previously done to the Soviet Union. Ultimately, therefore, Russia needs to develop similar capabilities if it is to defend itself and counteract Western aggression.

This highly politicised narrative originates from two different sources. However, before analysing this politicisation process, it is important to emphasise that those involved in this debate have been genuinely concerned with an attempt to prevent the repetition of Moscow's defeat in the Cold War. In surveys conducted since the dissolution of the Soviet Union, the Russian population has consistently expressed regret at its dissolution (66 per cent in 1992, peaking in 2000 with 75 per cent, and falling to 56 per cent in 2016), with as much as a third of the population believing that the fall of the USSR could have been avoided (with the highest percentages of 33 per cent in 2011 and 2016).[46] Given this context, it is not surprising that Russian academic and military experts have sought to learn the lessons of that defeat, which, in their opinion, was an outcome of the 'employment of psychological operations and special actions' by the West, which ultimately led to:

> The tragic consequences comparable only with a world war: the Soviet Union was destroyed, millions of people were affected, the living standards and morals fell sharply, the prospects of people and the state were threatened, [and] a part of the former Soviet republics sank down to the level of third world countries.[47]

The first source that has been politicising the narratives of what has come to be called the West's *gibridnaya voyna* is Russian academia. As was the case with politicisation of the narrative of Russia's hybrid/information/cyber warfare against the West by Western academics since the beginning of the Ukrainian Crisis in 2014,[48] Russian academia has promoted an opposing narrative since the late 1990s. In much the same way as Freedman's observations, discussed in the previous chapter, apply to the relationship between academics and policy-makers in the West,[49] the same also appears to have been the case in Russia. In other words, the Russian academic discourse on *gibridnaya voyna* (like the academic discourse on Russian hybrid warfare in the West) has become what Freedman describes as 'an echo chamber' that favours academics who share the views of officials because such academics 'are more likely to get a hearing'.[50]

The second major source of this politicisation is the Russian military. Unlike their Western counterparts, and especially NATO, which started to rethink their conceptual approach towards for non-military means and methods only after 2014,[51] Russian military thinkers have been focusing on the same problem since the early 2000s. As a consequence, it is not surprising that one of the main conceptual dilemmas in this field—how military organisations respond to threats that are non-military in nature—has been accurately addressed by Russian military thinkers. Whereas NATO has been forced to jump through various conceptual hoops in an attempt to counteract the full spectrum of the challenges of Russian hybrid warfare, Russian thinkers have solved this problem by drawing a conceptual division between the theories of new-generation warfare and *gibridnaya voyna*. As discussed previously, according to Chekinov and Bogdanov, non-military means and methods play an important part in both types of political confrontation; however, new-generation warfare has to be directed by the armed forces, while *gibridnaya voyna* should be conducted by the civilian leadership, indirectly supported by the military.[52]

By defining new-generation warfare within the traditional conceptual framework of the phenomenon of war, Chekinov and Bogdanov have sought to isolate war-making from policy-making. Within the context of the ongoing conceptual debate that has been trying to blend the two (i.e. information, net-centric, subversion, *gibridnaya voyna* and other theories that applied the term 'war' to something that is not war), they state that:

> Despite the arguments of some military scholars that there is a need for a radical revision of the essence and the content of war and that military power has lost its priority; the use of armed forces remains an obligatory criterion that distinguishes war, as a special period in the interstate confrontation, from other periods in the existence of states, thereby confirming the immutability of the approach to the essence of war.[53]

In other words, they maintain that while 'political–diplomatic, information, economic and other means and methods of influence on the opposing side are crucial in the prevention of conflicts and wars', and even have 'a significant impact on the character of the armed struggle', a new-generation war, regardless of its non-military elements, still requires a sophisticated and powerful military as 'the main characteristic of war is defined by the use of armed forces [and] the acts of violence'.[54] Understanding this conceptual division between the theory of new-generation warfare and the idea of *gibridnaya voyna* is highly important as it helps to point to the Russian military's political motives in promoting the theory of new-generation warfare as a new conceptualisation of modern conflict.

Analysing Chekinov and Bogdanov's works, as well as the articles published by General Gerasimov, many Western experts and analysts have pointed to the authors' emphasis on non-military means and methods in their conceptualisation of the theory of new-generation warfare.[55] However, a closer examination of these works reveals that, in fact, Russian military thinkers have placed an equal (or even greater) emphasis on the importance of military methods and actions, calling for 'the Russian Federation armed forces to be able ... to meet the challenges of the new-generation war'.[56]

For example, an analysis of Chekinov and Bogdanov's famous 'The Nature and the Content of the New-Generation War',[57] clearly shows the authors' focus on the importance of military means and methods in contemporary conflicts. Although they do indeed claim that non-military actions will be an integral part of new-generation warfare in 'assisting to weaken and eliminate military threats',[58] they consistently argue that, without the employment of armed forces, 'the achievement of the new-generation war aims will be impossible'.[59] They support this argument by dividing new-generation war into two main phases. The initial phase includes:

> a special information operation; a radio-electronic operation; an air-space operation; the systematic actions of the air force; [and] fire operation (based on the employment of precise weapons from different platforms, long-range artillery systems and weapons based on new physical principles) targeting the enemy in all directions covering the full depth of its territory.

This should then be followed by a second, concluding phase, consisting of:

> special operations conducted by reconnaissance units to locate enemy surviving units and communicate their coordinates to the [different] means of fire; contactless fire operations based on the newest effective means of destruction and intended to finalise the destruction of the resisting military formations; airborne operations that localise the resistance pockets; [and] the operations of land forces that cleanse the territory.[60]

In other words, new-generation warfare is not necessarily about non-military means and methods, but rather about the complex employment of armed forces, which is supported by indirect, non-military actions that 'create chaos and uncontrollability, demoralising people and the personnel of the defending military', thereby offering the aggressor an opportunity 'to achieve the desired military–political and economic aims of a military campaign in a very short period of time and without significant casualties'.[61] And if a new-generation war ultimately ends with a military campaign, it is not surprising that Chekinov and Bogdanov claim that while 'in the past decades the military–economic potential of Russia has been weakened', a new-generation war 'will be conducted according to the rules and laws of a country, which is most prepared to implement in practice the most progressive achievements in the fields of military–economy and technology'. In their view, military–technical superiority can only be achieved through: 'A continuous modernisation of a country's military–technical foundation, the creation of a highly effective economy, [and] the development and implementation of plans for the modernisation and production of weapons and military systems based on the newest technologies.'[62]

As a result, the theory of new-generation war has played two very important roles in Russian military discourse. On the one hand, it has institutionalised the role of non-military means either by isolating them as an independent, non-violent type of policy-making (i.e. *gibridnaya voyna*), or by integrating them as an essential, but secondary and preceding part of war-making. On the other hand, within the context of the ongoing conceptual debate (i.e. the theories of information, net-centric, subversion wars and others) that has tried to shift the centre of gravity towards non-military indirect actions, the politicisation of the concept of new-generation war has sought to preserve the status of military means and methods, ultimately calling for additional investment in the Russian armed forces.

Another example of this process is a 2013 article by the Russian chief of the General Staff, General Valery Gerasimov. As with Chekinov and Bogdanov's work, many Western experts came to associate this article with the idea of non-military means, and it has even been referred to as the 'Gerasimov Doctrine';[63] however, a closer examination clearly demonstrates that this is not the case. Thus, on the one hand, repeating the ideas expressed by Chekinov and Bogdanov, Gerasimov states that 'the role of non-military methods in achieving political and strategic goals has increased, [and] in some cases their effectiveness has significantly surpassed military force'. Yet, on the other, and again repeating the main narrative introduced in the theory of new-generation war, he argues that a demand for comprehensive modern weapons is an essential part of the preparation for future wars:

> An additional factor, which influences the transformation of the content of contemporary military struggle, is the employment of modern military robotic

complexes and research in the field of artificial intelligence. In addition to the already flying unmanned aerial vehicles, the battlefield of tomorrow will be filled by walking, crawling, jumping and flying robots. In a nearby future, it will be possible to create fully robotic formations able to independently conduct military actions.[64]

Hence Gerasimov essentially repeats the claim of Chekinov and Bogdanov that non-military means play an important part in contemporary conflicts, but that new-generation war is a war, and therefore it requires a significant investment in the armed forces.

Moreover, in a later article from 2016, Gerasimov further developed this idea, confirming the conceptual division between *gibridnaya voyna* and new-generation war, as well as the importance of the development and modernisation of Russian military power. Analysing the role and place of hybrid methods (*gibridnye metody*) as 'a complex employment of political, economic, information and other non-military means, which are supported by military power', Gerasimov concludes that: 'Today, a combination of traditional and hybrid methods is already an important characteristic of any armed conflict. However, if the latter can be used without an open employment of military force, the former—traditional military actions—cannot without hybrid [methods].'[65]

As already noted, this conceptual division between *gibridnaya voyna*, and its emphasis on non-military means, and the military-oriented new-generation warfare—which uses *gibridnye metody* as part of the traditional phenomenon of war—is vital from a political perspective, as it allows the authors to emphasise the importance of investing in the armed forces. According to Gerasimov: 'Regardless of the increasing importance of the non-military means in the resolution of interstate confrontations, the role of the armed forces, in providing the security of a state, is not decreasing, but only growing. Therefore, the requirements from the armed forces capabilities are [also] extending.'[66]

By conceptualising political, economic, information, ideological and other subversive non-military means and methods as part of contemporary military confrontation, Russian military thinkers have politicised the idea of new-generation warfare as a new type of armed conflict, which requires sufficient investment in the development and modernisation of the Russian military.

The politicisation of the idea of new-generation warfare as a new vision of contemporary conflict has undoubtedly been but one of the many ways in which the Russian military has fought to increase its funding. And it would certainly appear that this fight has been successful, as the share of military expenditure as a percentage of Russian GDP increased from 3.6 per cent in 2005 to 5.4 per cent in 2015.[67] In 2016, the Swedish Defence Research Agency concluded that:

> Russian Armed Forces are developing from a force primarily designed for handling internal disorder and conflicts in the area of the former Soviet Union towards a structure configured for large-scale operations also beyond that area. The Armed Forces can defend Russia from foreign aggression in 2016 better than they could in 2013. They are a stronger instrument of coercion than before.[68]

Similarly, in his analysis of the nature and impact of the exercises the Russian military conducted between 2011 and 2014, Johan Norberg argues that:

> After the exercise cycles in 2011–2014, in 2015 Russia's Armed Forces were most likely capable of launching large-scale conventional high-intensity offensive joint inter-service operations with support from other government agencies ... Furthermore, the involvement of other agencies and civil authorities at national and regional level suggests an approach to war as a society-wide effort.[69]

And since, in Gerasimov's vision, successful defence in the era of new-generation war, and especially its non-military elements, 'can be organised only by the complex employment of all the enforcement agencies of a state',[70] it would appear that his theory has found much support across the Russian political leadership, in addition to the military. Given the significant rise in the Russian defence budget,[71] it is important to examine how and why the narratives of *gibridnaya voyna* have been promoted by Russian politicians.

The politicisation of gibridnaya voyna *and its narratives in Russian political discourse*

In the Russian political discourse, it is difficult to find direct references to terms such as net-centric or subversion wars, *gibridnaya voyna* or *gibridnye metody*. However, as Andrew Monaghan points out, the main narrative underlying these theories (i.e. the West's use of non-military means to subvert and undermine the Russian political regime) is a dominant theme in Russian political discourse.[72] A closer examination of this discourse suggests that the conceptual framework Russian politicians most often evoke is that of 'information warfare'. For example, when discussing the Western reaction to Russia's actions in Crimea and eastern Ukraine, President Vladimir Putin claimed that:

> Our diplomats understand, of course, how important the battle to influence public opinion and shape the public mood is these days. We have given these

issues much attention over recent years. However, today, as we face a growing barrage of information attacks unleashed against Russia by some of our so-called partners, we need to make even greater efforts in this direction.[73]

Another example is the Russian Minister of Foreign Affairs, Sergey Lavrov, who, elaborating on the nature of these attacks, argues:

> The destructive [political] line related to the events in Ukraine, the introduction of the illegitimate sanctions against Russia, the attempts to punish our country for upholding truth and justice, for speaking in defence of [our] compatriots [in Ukraine] … [all these] led to a serious crisis in our relations with the West. We are faced with a large-scale information war.[74]

Interpreting the nature of the Western sanctions against Russia, Lavrov went on to say:

> The leaders of the Western countries publicly declare that it is necessary to ensure that the sanctions destroy the economy and raise popular protests … the West clearly demonstrates that it does not merely seek to change Russia's policy … but [it] wants to change the regime. … [The United States] uses financial and economic pressure, information attacks … and, of course, informational and ideological influence, supported by externally funded non-governmental organisations.[75]

The very same narrative has also been expressed and promoted by Putin:

> We also face attempts to use the so-called 'colour technologies'—from organising unlawful street protests to open propaganda of hostility and hatred in social networks. And the goal is obvious—to provoke civil conflicts, to strike a blow at the constitutional foundations of our state, [and] at the sovereignty of the country.[76]

While these arguments clearly repeat the narratives of *gibridnaya voyna* and were expressed in the context of the Ukrainian Crisis, it is clearly the case that the Russian political establishment has been promoting the idea that the West had been seeking to undermine the Russian political establishment long before the beginning of the crisis. For example, in his speech at a meeting with ambassadors and permanent representatives of the Russian Federation in 2006, which was in many respects the domestic version of the 2007 speech at the Munich Security Conference, Putin had already stated that:

> After the destruction of the bipolar order, the unpredictability in global development has remained strong. Perhaps this is the reason for the unceasing talks about the inevitable conflict of civilisations with a prospect of prolonged confrontation that would follow the example of the 'cold war' … not all were prepared for the fact that Russia would begin to regain economic health and its position on the world

stage so quickly. Someone looks at us through the prism of the prejudices of the past ... and sees a threat in a strong, [and] resurging Russia.[77]

Since, in the Kremlin's view, the West has been using different non-military actions to undermine 'a strong and resurging' Russia through the information dimension, it is easy to trace the politicisation of this argument via a comparative analysis of different Russian conceptual documents. A comparison of the 2000 and 2016 Information Security Doctrines, for example, offers a very interesting picture of the development in the Kremlin's understanding of the information dimension as a threat to the Russian state. One of the main differences between these two doctrines is that, while the 2000 doctrine divides threats to Russia's information security between external threats and internal ones, the 2016 doctrine makes no such distinction.

The external threats listed in the 2000 version of the doctrine understandably include the 'subversive activity of foreign special forces based on methods of information–psychological influence' and 'the activities of foreign political, economic and military institutions directed against the interests of the Russian Federation'. Yet, at the same time, the document views the following threats as internal:

> The activity of public organisations aimed for a violent change of the constitutional system's foundations and the violation of the integrity of the Russian Federation, the incitement of social, racial, national and religious hostility, [and] the dissemination of these ideas in the mass media;
>
> The information and propaganda activities of political powers, public organisations, mass media outlets and individuals that distorts the strategy and tactics of the foreign policy activities of the Russian Federation;
>
> Possible information and propaganda activities that undermine the prestige of the Armed Forces of the Russian Federation and their combat readiness.[78]

Thus, in 2000, the Russian government not only recognised a very clear distinction between external and internal threats but also acknowledged, de facto, that different actions that undermine the Russian government could occur independently of one another due to different internal domestic factors, regardless of external subversive actions.

In contrast, the 2016 version of the doctrine does not draw this distinction, and instead emphasises the causal connection between external influence and the internal destabilisation of the state:

> The special services of individual states increase their employment of the means of information–psychological influence aimed to destabilise domestic political

and social situation in various regions of the world, leading to a subversion of sovereignty and a violation of the territorial integrity of other states;

The state of information security in the field of a country's defence is characterised by an increase in the extent to which individual states and organisations use information technologies for military–political purposes, including actions that contradict international law and are aimed at undermining sovereignty, political and social stability, the territorial integrity of the Russian Federation and its Allies, and pose a threat to international peace and global, and regional, security.[79]

A similar amplification of the role of external influence in domestic affairs can also be traced throughout different versions of Russia's Military Doctrine. In its discussion of the nature of contemporary conflicts, the 2000 version simply argued that they are characterised by 'an active information struggle [and] a disorientation of public opinion in individual states, as well as global population', and by 'the desire of the actors to disorganise the state and military control systems'.[80] The 2010 version advances this idea, stating that contemporary conflicts are characterised by 'a complex employment of military power and non-military means and methods' and the use of 'information struggle actions in advance, aimed to achieve political goals without a deployment of military force'.[81]

The 2014 version of the doctrine takes this amplification even further, arguing that contemporary conflicts involve 'a complex employment of military power, political, economic, informational and other non-military means, implemented with a widespread use of the protest potential of the population and special operations efforts', as well as 'the use of externally funded and controlled political forces [and] public movements'.[82] Moreover, referring directly to the threats posed by the information dimension, the 2014 doctrine states that one of the threats to the security of the Russian Federation includes 'actions of information influence on population, first and foremost on young citizens, aimed to subvert historical, spiritual and patriotic traditions'.[83]

The same narrative can also be traced across the different versions of Russia's Foreign Policy Concept. When discussing the state of the contemporary world, all four latest versions (2000, 2008, 2013 and 2016) state that, in addition to military power, other non-military factors have become equally effective as a means of influencing international affairs. However, whereas the 2000 and 2008 versions argue that this offers an opportunity 'for building a more stable and crisis-resistant world order,'[84] and the 2013 version states that 'economic interdependence between states is one of the key factors that supports international stability',[85] the 2016 version takes an entirely opposite approach, arguing that the existence of effective non-military means in international relations creates 'a desire to employ corresponding opportunities for a realisation of geopolitical interests, which harms the search for ways to resolve disputes and peacefully solve existing international problems, basing it on the norm of international law'.[86]

This analysis of the doctrinal documents demonstrates the shift in the Kremlin's approach towards the role and place of non-military means and methods in international relations and the slow, but consistent, politicisation of external, non-military factors (i.e. political, economic, informational) as the main threat to the stability of the Russian government. As stated in the 2015 Russian Nation Security Strategy:

> The growing confrontation in the global information space has an increasing influence on the character of the international situation, as an outcome of the desire of some countries to achieve their geopolitical objectives by using information and communication technologies, including the manipulation of public consciousness and the falsification of history.[87]

Any attempt to explain the reason behind this politicisation needs to recognise that, from the Russian perspective, the dissolution of the Soviet Union resulted from the information–psychological operations and other non-military means and methods conducted by the West. In the mid-2000s, when a wave of economic success and growing domestic support had allowed the Russian leadership to start reviving its position in global affairs,[88] the Kremlin also started to realise that the West would not accept the Russia's revived geopolitical interests. As Putin recalled in the documentary *President*:

> I served in the KGB, in foreign intelligence, for almost 20 years, and even I thought that with the collapse of this ideological obstacle, in the shape of the Communist Party's monopoly on power, everything will fundamentally change. No ... it did not change in essence, because, as it turned out ... there are geopolitical interests that are completely disconnected from any ideology. This is where our partners [the West] were required to understand that Russia is a country that has, and it cannot be without, its own geopolitical interests. And it is required to respect each other, while searching for balances and mutually accepted resolutions.[89]

Since the beginning of the second decade of the twenty-first century, and especially since the 2012 presidential elections, which involved unprecedented public demonstrations in opposition to the ruling regime,[90] the Kremlin has started to feel threatened by 'an employment of information technologies aimed to harm the sovereignty, territorial integrity, [and] political and social stability of the Russian Federation.'[91] Therefore, it is not surprising that the Kremlin, in an attempt to avoid the repetition of the Cold War's outcome, came to the conclusion that its political stability can only be achieved by controlling and restricting the influence of external institutions (such as the mass media and NGOs) on the Russian public.

The politicisation of the narrative of Western non-military intervention in Russian domestic affairs since the early 2010s in political rhetoric and doctrinal documents was aimed at preparing the ground for corresponding legislation intended to minimise the perceived threat of external influence on Russian society. In the past several years, the Russian government has passed several important pieces of legislation with the intention of counteracting the alleged Western non-military offensive and protecting the Russian information space, such as Federal Law no. 121-FZ, passed in 2012, which restricts the activity of NGOs that receive foreign funding;[92] Federal Law no. 398-FZ (2013), which simplifies the procedures required to block extremist websites; Federal Law no. 97-FZ (2014), which enforces governmental supervision over websites and blogs; Federal Law no 374-FZ (2016), which forces websites to store the data of their Russian clients in Russia;[93] and Federal Law no. 305-FZ (2014), which limits foreign investment in Russian media outlets.[94] Moreover, it seems that the demonisation of the West's information offensive against Russia, which preceded these laws, has served the Kremlin well: while this new legislation was followed by a wave of critical reaction internationally and domestically,[95] this criticism has had very limited consequences and the laws have been widely accepted by the general public.

Conclusions: 'gibridnaya voyna *against us is coming, it has already begun!*'

As discussed throughout this chapter, the Russian discourse on the Western *gibridnaya voyna* has two main interconnected elements: military and political. On the military side, the rise of the theory of new-generation warfare played a pivotal part in institutionalising the idea of *gibridnaya voyna* as a defined type of international confrontation.

On the one hand, Russian military thinkers have tended to agree that the role of non-military actions is increasing. Their analysis of the Cold War and the collapse of the USSR, as well as their observations of new information technologies and globalisation processes, has led them to accept that the twenty-first century offers means and methods for the non-violent resolution of certain inter-state confrontations, which could only have been solved previously through the use of war. On the other, in the context of the conceptual debate on the rising importance of non-military means, Russian military thinkers and strategists have come to the conclusion that this emphasis on non-military means and the subsiding status of war as an armed struggle, would also mean the declining importance of armed forces. They responded to this situation by developing the concept of new-generation war, which institutionalised non-military means and methods as a part of the traditional definition of war.

As discussed above, the development of new-generation war theory was a logical step in Russian military thought in the context of the Russian conceptual and theoretical debate over the non-military elements of war since the late 1990s. Neither Chekinov and Bogdanov, nor Gerasimov, were true pioneers in suggesting that non-military indirect actions play an important role in contemporary conflicts. Yet their theory of new-generation war has been vital in the internal Russian debate over the nature of inter-state confrontations in general, and the role of the armed forces in particular.

The conceptual division between new-generation war, as an armed struggle accompanied by non-military actions, and *gibridnaya voyna*, as a non-violent type of political confrontation based solely on non-military means, was a crucial step in Russian military thinking. As this chapter has shown, the theory of new-generation warfare offers very little that is new in terms of non-military, indirect means and methods. Its main contribution, however, is in its emphasis on the importance of military means and capabilities, which created the conceptual–theoretical foundations required to revive the status of the armed forces, and, therefore, their funding.

The idea of *gibridnaya voyna* within Russian military discourse seems to be a by-product of the new-generation war conceptualisation, simply because the definition of the latter (as an armed struggle accompanied by non-military means) required a title for the former (as a struggle that solely constitutes non-military means). Russian military thinkers strongly disagree with the semantics of the title, arguing that *gibridnaya voyna* is not *voyna* (war) in its essence because it does not involve the use of military force. As two Russian officers put it: '*Gibridnaya voyna* is an operation intended to tear the territory of another state apart, [and it is] based on a coordinated complex application of political–diplomatic, information–propagandistic, financial–economic, as well as military, measures. At the same time, the military campaign itself is not conducted.'[96]

'GIBRIDNAYA VOYNA AGAINST US IS COMING'

Yet it would seem that they generally tend to accept this concept, because an institutionalisation of a non-military struggle, regardless of its title, is vital for the institutionalisation of new-generation war. Therefore, the Russian military has been promoting the idea of *gibridnaya voyna* and its narratives as a by-product of their promotion of the theory of new-generation war. And, since the politicisation of the latter is important to attract the funding required to face the technological challenges of twenty-first-century warfare, the politicisation of the former helps to achieve this.

The reason for the Russian military's politicisation of the new-generation war concept becomes even clearer in the context of the Russian political discourse on the role of non-military threats. As discussed above, since the mid-2000s, and especially after the 2012 presidential elections, the Kremlin has been claiming that the West's subversive, non-military actions against Russia pose the main threat to Russia's integrity and stability. Moreover, it seems that the Kremlin has found it expedient to promote the ideas expressed by *gibridnaya voyna* as the main narrative surrounding the evil West and its bid to secure global domination.

The semantic similarities between *gibridnaya voyna* and its Western counterpart, hybrid war, used in the Western discourse to describe Russian actions, have allowed Russian politicians to accuse the West of employing *gibridnaya voyna* methods. As Sergey Lavrov put it:

> It has become fashionable to discuss that Russia is waging some kind of 'hybrid war' in Crimea and Ukraine. It is a very interesting term, but I would apply it, first and foremost, on the U.S and American war strategy—it is truly hybrid and it is aimed less at the military destruction of the enemy, but rather more at a regime change in states that lead policies which contradict Washington's line.[97]

However, at the same time, the politicisation of the Western *gibridnaya voyna* and its subversive methods has helped the Russian political leadership to close ranks against a new/old enemy, the West, mobilising Russian public opinion and the support of other Russian major political actors. In the words of Gennady Zyuganov, the leader of the Russian Communist Party:

> We have to understand that we were defeated neither by Hitler, nor by Napoleon. We were defeated by anti-Sovietism and Russophobia, and today these flames are ignited again ... We cannot be negligent, I experienced this in the 90s ... We live in an extraordinary situation, *Gibridnaya voyna* against us is coming, it has already begun ... It presents a colossal danger, and we have no right to relax.[98]

Moreover, when analysing Russian public opinion on the Western *gibridnaya voyna* against Russia, this politicisation has clearly had the desired effect.

The evidence of this success is that the controversial legislation intended to defend Russia against the Western *gibridnaya voyna* was generally supported by the Russian public without any significant objections.

This lack of criticism might be explained by the public's acceptance of the narratives surrounding the West's non-military, subversive offensive against Russia. In research carried out by the Levada Centre in December 2014, 87 per cent of respondents said that the West is hostile towards Russia, and 42 per cent of them claimed that this hostility was expressed in the form of an information war. The figures remained high in a similar survey conducted in October 2015, with 82 per cent stating a belief that the West was hostile towards Russia, and with 44 per cent of respondents accusing the West of waging an information war against the country. Interestingly, in their answers to the question 'What does the West try to achieve by toughening the sanctions against Russia?' the vast majority of respondents answered 'To weaken and humiliate Russia' (71 per cent in September 2014 and 69 per cent in October 2015), in contrast to only 4 per cent (in 2014, and 6 per cent in 2015) who believed that by tightening sanctions the West was trying 'to stop the war, destruction and people's deaths in Eastern Ukraine'.[99] While it seems that Russians truly believe in the narrative of the Western offensive to undermine Russia, the possibility cannot be excluded that these figures might be an outcome of 'the Kremlin's propaganda machine [that] goes further in manipulating Russian mass public opinion'.[100]

It is important to remember, however, that the grief for lost honour and national pride after the fall of the Soviet Union might also explain the thirst of the Russian public for a sound explanation of their defeat to Russia's traditional enemy—the West. And since the idea of *gibridnaya voyna* offers this explanation, it may simply be the case that the hostile attitudes of many Russian people towards the West are not the outcome of the Kremlin's brainwashing machine, but rather a result of what they truly hold in their hearts and minds—or even a genuine combination of both.

Either way, it seems that, not only in the minds of the Russian leaders but also in the hearts of Russian people, '*Gibridnaya voyna* against us is coming, it has already begun.'

CONCLUSIONS

THE RISE OF RUSSIAN 'HYBRID WAR'—LESSONS FOR THE WEST

As promised in the introduction, this book was not about the practicalities of the contemporary information struggle between the Kremlin and the West. It did not discuss the tactics of what Western experts commonly describe as Russian hybrid, cyber or information wars with their hacks, disinformation and propaganda operations conducted by pro-Russian forces, or forces backed by Russia.[1] Instead, this book has focused on the development of the conceptual discourse on the theory of contemporary conflicts in Russia and the West, as well as how each side has been interpreting the concepts of another. As the analysis of the discourses discussed throughout the book shows, in the case of both Russia and the West these discourses were infected by the attempts of different actors to promote their own political interests, rather than a more composed, dispassionate and professional attempt to understand the true nature of conflicts in the contemporary geopolitical environment and the state of modern technology. This understanding, however, seems to be of paramount importance, as it provides a conceptual strategic framework that allows us not only to comprehend the strategy of the adversary, his 'direction and use made of means by chosen ways in order to achieve desired goals',[2] but also to counter it in the most efficient and beneficial way. In other words, to successfully respond to the challenges contemporary Russia poses, whether they are hybrid, information, cyber, or any other voguish title, the West needs to formulate an apolitical vision of contemporary conflicts in general, and particularly within the context of contemporary Russia. This book's focus on the discourse surrounding this vision provides several important insights for the West.

Lesson one: on the nature of 'hybrid' and the danger of 'war'

This book's analysis of the discourse on hybrid warfare reveals that three completely different phenomena have been described as hybrid war. The first is the original conceptualisation given by Hoffman, according to which hybrid war 'incorporates a range of different modes of warfare including conventional capabilities, irregular tactics and formations, terrorist acts including indiscriminate violence and coercion, and criminal disorder'.[3] As discussed in Part 1, this theory was rooted in the conceptual debate over the nature of contemporary conflicts that had been occurring in the United States since the early 2000s, borrowing different elements from previously developed theories, such as unrestricted warfare, compound warfare and fourth generation warfare, as well as different observations outlined in the '2005 National Defence Strategy'. Hoffman's original theory, based on the example of the Second Lebanon War, was intended to describe the changing nature of contemporary conflicts, focusing on the tactical and operational levels. Therefore, the main contribution of his theory resides in its attempt to 'better prepare for that messy grey phenomenon and avoid the Groznys, Mogadishus and Bint-Jbeils',[4] rather than to suggest 'a strategic plan for an Army institution that is posturing itself for the long term'.[5] Nevertheless, this conceptualisation has been criticised for its 'conceptual weaknesses [which] serve as an impediment to clear and productive strategy making'.[6]

This criticism was one of the reasons for the re-conceptualisation of hybrid warfare in the early 2010s, leading to a broader definition of hybridity in contemporary conflicts as 'a complex blend of means that includes the orchestration of diplomacy, political interaction, humanitarian aid, social pressures, economic development, savvy use of the media, and military force'.[7] By 2012, NATO had generally abandoned its earlier work on hybrid warfare and hybrid threats, yet the Ukrainian Crisis and Russian actions in Crimea and eastern Ukraine served as a trigger that brought the idea of hybrid warfare back to the centre of the academic, military and policy-relevant debate. In an attempt to understand the Kremlin's surprisingly effective actions, which did not fit within any of the Western conceptual boxes on contemporary conflicts, NATO revived the concept of hybrid war, giving it an even broader definition than before: 'a continuum of threat, including unconventional and conventional methods'[8] that 'bridges the divide between the hard and soft power'.[9]

CONCLUSIONS

In tandem with this development, an independent discourse on hybrid warfare (*gibridnaya voyna*) had been developing in Russia. As discussed in Chapters 3, 4, 5 and 7, during the late 1990s and early 2000s, Russian academics and military thinkers were preoccupied with theorising the role of non-military means and methods in contemporary conflicts. While the term itself first appeared in Russian discourse only after 2014, as a direct response of the Russian academic community to the politicisation of hybrid warfare in the West as something that 'Russia allegedly wages in Ukraine',[10] the Russian conceptualisation of *gibridnaya voyna* was based on existing theories in Russia such as subversion-war, net-centric war and information war. Moreover, while Hoffman's hybrid warfare and 'Russian hybrid warfare', as developed by NATO, present different, though conceptually interconnected views on contemporary conflicts, the idea of *gibridnaya voyna* represents something entirely different, as its purpose is 'to achieve political goals with a minimal military influence on the enemy ... by undermining its military and economic potential by information and psychological pressure, the active support of the internal opposition, partisan and subversive methods'.[11]

Hence three different interpretations of hybrid warfare have come to dominate contemporary academic and military discourse: the original definition of hybrid warfare, which implies a combination of conventional and irregular means and methods on the battlefield; so-called Russian hybrid warfare, which includes any possible combination of military and non-military means; and *gibridnaya voyna*, which is solely based on non-military means and is intended to undermine and subvert an adversary without recourse to military force. Given these differences, the contemporary conceptualisation of hybridity is more a mélange of different concepts than a theory that helps to understand and interpret the complex reality of contemporary conflicts. There are two reasons for this: (1) the ambiguous nature of the term 'hybrid', and (2) the misuse of the term 'war'.

According to its linguistic definition, 'hybrid' means 'something that consists of or comes from a mixture of two or more other things'.[12] Hence while the term 'hybrid' implies that the phenomenon that it characterises consists of different elements, it specifies neither the nature of these elements, nor their proportions. The nature of the mixed elements is usually provided by the definition of the phenomenon: 'hybrid diesel-electric engine', for example, means that the engine is partly diesel and partly electric. Following this logic, it can be argued that Hoffman's concept of hybrid warfare is partly based on the use of conventional and partly irregular means, methods and technologies; Russian hybrid warfare, conversely, is partly based on military (partly covert and partly overt) operations and partly non-military (partly informational, partly cyber, partly diplomatic, etc.) means and methods; and *gibridnaya voyna* is partly economic, partly informational, partly diplomatic, partly cyber, and so on. Yet unlike the 'hybrid diesel-electric engine', which consists of two clearly defined parts, the variety of possible elements and their possible contributions to any of these forms of warfare creates a situation where the term hybrid warfare 'loses its value and causes confusion instead of clarifying the "reality" of modern warfare'.[13] Hence, a description of warfare as 'hybrid' implies any possible combination of all possible means and methods, which is of little help if we want to understand the true nature of contemporary conflict.

This confusion is best demonstrated by an example from the art world. While a combination (i.e. hybrid) of two colours, yellow and blue, can be described as a 'hybrid yellow-blue colour', this term completely fails to describe the whole pattern of possible colours produced by combining (hybridising) yellow and blue, such as 'green', 'lime', 'spring bud', 'tropical rain forest' or the many other colours that can be created by mixing blue and yellow in different proportions. Consequently, describing war as 'hybrid' is similar to describing grass as a 'hybrid yellow-blue colour', and not as 'green'—it is conceptually confusing and ultimately unhelpful for clarifying the nature of the described phenomenon.

In much the same way as all possible colours of grass fall under the definition of a hybrid between yellow and blue, while not all possible hybrid colours of yellow and blue can be used to describe grass, any war is a hybrid of military and non-military means and methods, but not any hybrid of these elements can be described as war. The misuse of the term 'war' will be discussed below; here, it is simply important to reflect on the conceptual ambiguity created by the implications of the term hybrid for the Ukrainian Crisis.

As discussed above, the West has accused Russia of employing hybrid warfare in Ukraine, whether in its original theoretical guise as formulated by Hoffman or in the reconceptualised one. Russia, in turn, has accused the West of waging *gibridnaya voyna* against it. As these three types of confrontations are completely different, to some extent both the West and Russia can be said to be correct. First, as the separatist movements in eastern Ukraine supported by Russia employ a combination of conventional and irregular tactics, technologies and methods of warfare, they clearly present a hybrid threat, according to Hoffman's original definition, to the Ukrainian military. Secondly, the combination of military and non-military, covert and overt means and methods that the Kremlin employed in Crimea is evidently consistent with the definition of Russian hybrid warfare as conceptualised by NATO in 2014–15. Finally, the West's use of non-military, subversive means and methods against the USSR during the Cold War,[14] which it has continued using to extend its influence in the post-Soviet space, thus undermining Russian geopolitical interests,[15] falls under the definition of *gibridnaya voyna*.

The term hybrid can therefore be used to fit any of these interpretations, as each of them represents a certain combination of different means and methods employed to achieve political goals. Yet the effectiveness of this term to describe specific combinations in an attempt to counteract them is similar to applying the term 'hybrid yellow-blue colour' to describe the grass in somebody's garden. Therefore, it is not surprising that the terminology of hybrid warfare has generally been rejected by military organisations, both in the West and in Russia, in favour of more specific definitions. It is similarly unsurprising that the term hybrid war has been so quickly and successfully politicised in the West and Russia, as almost any hostile action could be described under the same conceptual rubric, thereby creating a continuity of unified political messaging and allowing different domestic actors to close ranks against an external threat.

Despite the political usefulness of the term hybrid warfare, it would appear that Russian and Western military professionals now recognise that the term is next to useless for describing the real nature of contemporary conflicts, leading them to promote more specific definitions, such as information warfare and cyber warfare, which are now prevalent in the West, or new-generation war, which is currently prevalent in Russia. Although each of these new terms presents a certain hybrid (combination) of different means and methods, the real question is whether all of them can be defined as war. In other words, if, as Clausewitz put it, 'there is only one means of war: combat',[16] it would be logical to conclude that a hybrid of hostile means and methods without combat is not an actual war. While the famous Clausewitzian dictum that 'war is merely the continuation of policy by other means'[17] could be interpreted as a suggestion that a war can be conducted without military force, he also makes it clear that 'war ... is an act of force to compel our enemy to do our will'.[18] Moreover, specifically referring to the relations between politics and war, and the element required to transform hostile diplomacy into war, Clausewitz states: 'at the highest level the art of war turns into policy—but a policy conducted by fighting battles rather than by sending diplomatic notes'.[19]

Consequently, it is not surprising that, given the current tendency to describe non-military confrontations as war (i.e. information or cyber war, political war or *gibridnaya voyna*), some scholars have rightly argued that 'there are no wars in history that were won by non-military means, or by the use of information, alone',[20] because, without using military force, such confrontations are not wars, and to describe them as wars is 'a dangerous misuse of the word "war".'[21] While the danger of misusing the word war in contemporary political discourse has been widely discussed in both Russia[22] and the West,[23] General Makmut Gareev perhaps put it best, stating that 'if an employment of any non-military means is a war, then the whole of human history is war',[24] as the 'over-free employment of such a word as "war" devalues the severe [nature of this] concept and dulls its adequate perception in society'.[25]

On the one hand, there is no doubt that some of the concepts discussed in this book fully deserve the title of war. For example: unrestricted warfare, compound warfare, fourth generation warfare, Russian new-generation warfare, and even Hoffman's original conceptualisation of hybrid warfare involve some use of military force, and, therefore, represent phenomena that fall under the definition of war. On the other hand, however, Messner's theory of subversion-war, Panarin's theory of information war, Dugin's theory of net-centric war, together with *gibridnaya voyna*—as well as the Western interpretations of Russian information and cyber warfare—do not involve a violent clash of military forces intended to achieve political goals. As such, in the words of Chekinov and Bogdanov, these concepts describe actions that 'can be defined as anything else—"political conflict", "incident", "competition"—but not as a war'.[26]

Moreover, as has been discussed throughout this book, extending the definition of war to any possible combination of political confrontation is a dangerous exercise; though doing so might serve political ends, it is also very confusing for the military. Any military organisation, by definition, takes preparations to win wars (i.e. to compel the enemy through an act of force), and, therefore, conceptualising non-military confrontations as wars is unhelpful for the military leadership, simply because most of the required actions and counter-actions do not fall under the military's responsibility. This is why the Russian military has been so eager to draw a distinction between new-generation war and *gibridnaya voyna*, as well as other associated theories. This is also why NATO has been trying to counteract Russia's non-military actions in the Baltic states and in Eastern Europe by deploying the Very High Readiness Joint Task Force (VJTF)—as a military alliance, it can only employ military means and methods, even when facing a non-military challenge.

To conclude this book's discussion of the nature of hybrid war, with all its different interpretations and associated concepts, it is important to try to

CONCLUSIONS

make some sense or order out of the different conceptualisations and re-conceptualisations, as well the various interpretations and misinterpretations of the concepts, which have been occupying Western and Russian discourses on the nature of contemporary conflicts. Based on the definitions given throughout this book, it is possible to reach four conclusions. First, the theory of compound warfare, as proposed by Huber in the late 1990s, and the idea of hybrid warfare, as formulated by Hoffman in the mid-2000s, are similar in nature as both deal with the operational aspects of a combined employment of regular and irregular forces against a single enemy. The difference is that, while the former discusses the general concept, tracing its implications through historical cases, Hoffman's theory focuses on the specific characteristics of compound warfare in the twenty-first century, applying this concept to the techno-tactical and operational environments of the contemporary battlefield.

The second conclusion is that the concept of fourth generation warfare, as formulated by Lind, as well as Qiao Liang and Wang Xiangsui's theory of unrestricted warfare, and NATO's re-conceptualisation of hybrid warfare, together with the concept of new-generation war, formulated by Chekinov and Bogdanov, refer more or less to the same phenomenon—a mix of military force with indirect non-military means and methods, all combined together as one strategy in an attempt to achieve the political goals of an armed struggle. While none of these theories offers a strategy guaranteed to end in victory, they all emphasise the importance of combination in contemporary wars; as Qiao and Wang put it: 'whoever [will be] able to mix a tasty and unique cocktail for the future banquet of war will ultimately be able to wear the laurels of success on his own head'.[27]

The third conclusion relates to the different theories of political subversion that were promoted in Russia during the 1990s and 2000s, namely subversion-war, net-centric war, information war and so on, which ultimately led to the concept of *gibridnaya voyna*. Both the latter and the contemporary Western conceptualisation of Russian information or cyber warfare, which replaced the rejected concept of hybrid war, bear important similarities to the US concept of political warfare, developed during the Cold War, as all imply a certain use of 'violence, economic pressure, subversion, and diplomacy' where 'the use of words, images, and ideas, [is] commonly known ... as propaganda and psychological warfare'.[28]

Finally, that the word 'hybrid' appears in all of these different groups of concepts serves to demonstrate, yet again, the term's ambiguity when applied to the phenomenon of war, which, by definition, is already multifaceted and highly complex.

Lesson two: on the novelty of hybrid war

The second lesson concerns the novelty (or otherwise) of hybrid warfare in any of its three interpretations. In an attempt to attract wider attention, researchers often claim that their findings and theories are entirely new—there is, after all, no better way to draw attention to research on Russian actions during the Ukrainian Crisis than to proclaim that 'we have witnessed the application of a new type of warfare where dominance in the information field and hybrid, asymmetric warfare are the key elements'.[29] However, it is difficult to disagree with Freedman, who warns against using new terms for old concepts.[30] Indeed, as we have seen, the information field and propaganda have played a vital role in conflicts since antiquity,[31] and any form of warfare 'is to a degree asymmetric, because states rarely have identical capabilities or strategies',[32] and the term 'hybrid', as discussed above, can be used to describe more or less anything.

This may ultimately be the best explanation for the selective application of the term hybrid warfare to contemporary conflicts. For example, while the US-led wars in Afghanistan and Iraq could be described as hybrid, the theorists of hybrid warfare generally avoid doing so. There are two reasons for this. The first is that since these wars were waged against irregular threats, rather than a combination of regular and irregular threats, they do not fall under Hoffman's original definition of hybrid warfare. The second is that the development of a broader concept of hybridity was carried out within the specific political context of tension between Russia and the West, and the new concept was so broad—'Hybrid Warfare bridges the divide between the hard and soft power'[33]—that it could not be used to describe the adversaries in Iraq and Afghanistan as hybrid. Similarly, very few, if any, researchers have described the ongoing confrontation with ISIS as hybrid war, despite ISIS presenting a complex mix of regular and irregular capabilities, methods and techniques, together with a highly effective information campaign.[34]

The absence of analyses in the literature describing these wars as hybrid, despite their hybrid nature, further demonstrates the ambiguity of the concept itself, but also its political usefulness, because the decision to apply it explicitly to the confrontation between the West and Russia seems to be more political than conceptually appropriate. It also brings into question the ability of hybrid warfare to offer a new conceptual vision of contemporary conflict.

CONCLUSIONS

In discussing the potential novelty of any new concept of warfare, it is important to identify whether the concept attempts to describe a new battlefield reality (i.e. tactical and operational) or a new way of strategy-making. Since strategy is 'the relationship between political ends and (military, economic, political, etc.) means',[35] it is possible to paraphrase the metaphor used by the two Chinese colonels, arguing that strategy-making is a process of mixing a cocktail from these means, and 'the winner is the one who combined [them] well'.[36] Hence, a new concept that claims to contribute to strategy-making is required to produce a new recipe (either by offering a new cocktail, or by inventing a recipe for a cocktail that was previously mixed by intuition rather than by knowledge); a new concept on the tactical–operational level simply attempts to describe an improvement of ingredients mixed according to an old recipe. The best example of the first is the concept of strategy developed during the nineteenth century—since the 'cocktail' of 'political aims and the use of force, or its threat' can be traced back to antiquity,[37] the novelty of the nineteenth-century strategists was in inventing its recipe. The best example of the second type is, in fact, Hoffman's hybrid warfare.

From the analysis of the original concept of hybrid warfare in Part 1, it is clear that the concept attempts to describe a new tactical-operational environment rather than a new way of strategy-making, as has been demonstrated in the criticism of the concept's 'a-strategic' nature.[38] On the other hand, as Murray, Mansoor, Huber and others have demonstrated, there is little new in the recipe for the 'cocktail' that Hoffman's hybrid warfare deals with—a mix of regular and irregular forces, means and methods.[39] Yet, on the other, Hoffman updated this cocktail for the realities of the twenty-first century, describing the new 'taste' that new technologies, and subsequent new tactics and operations, have brought to the old strategy-making 'cocktail' of combining regular and irregular types of warfare. Moreover, in defence of the novelty of Hoffman's contribution, it might be argued that the differences in the equipment, weapons, training and skills between contemporary regular and irregular forces are greater than they used to be in the past, and therefore their mixture creates a truly new tactical-operational 'taste' for the old strategy-making cocktail.

From the discourse on the new interpretation of hybrid warfare as a mix of military and non-military means and methods (and all similar concepts discussed throughout this book, such as fourth-generation war, unrestricted warfare and new-generation war), it is clear that they all claim to offer a new vision of strategy-making. However, a closer examination of history suggests that strategy-making cocktails of military and non-military means and methods have existed since antiquity, as 'anything that might eat away at the enemy is considered worth trying'.[40] For example, France employed this strategic cocktail against Britain in the American Revolution by supporting the American colonies;[41] Germany did the same by funding the Bolsheviks against the Russian Empire during the First World War;[42] and Britain, fighting the Ottoman Empire, also employed a mix of diplomatic, economic and other military and non-military means in support of the Arab Revolt.[43] Since there is nothing new in this, the only novelty in the concept of new-generation war, as well as in all similar theories that blend military and non-military means, is their attempts to suggest how to mix this cocktail to fit the twenty-first-century environment. In other words, despite claiming that they offer a new vision of strategy-making, these concepts simply adjust a well-known recipe to the contemporary political, economic, technological realities, and therefore, they merely offer a new tactical-operational recipe for mixing the old strategy-making cocktail.

The same conclusion also applies also to the concept of *gibridnaya voyna* and all other associated theories that imply an employment of non-military means alone, as they all suggest nothing that is conceptually new in terms of strategy-making. The importance of non-military means for successful strategy-making has been known since the time of Sun Tzu, who stated that 'hence to fight and win all your battles is not the foremost excellence; to break the enemy's resistance without fighting is the foremost excellence'.[44] An analysis of history shows that non-military 'cocktails' based on 'words, images, and ideas ... [as well as] propaganda and psychological warfare' have been mixed for as long as humanity has existed.[45] The precise receipe of these cocktails has undoubtedly changed in response to different military, political, economic and social developments, as well as technological progress. However, as stated above, this type of adjustment is simply a tactical and operational novelty, rather than a new way of strategy-making. In other words, *gibridnaya voyna*, by its nature, is not a new phenomenon, nor exclusively Russian, but an adaptation of a very old and well-known practice to the geopolitical situation and technological capabilities of the twenty-first century.

Accordingly, neither Russia's new-generation warfare, nor the discourse surrounding *gibridnaya voyna*, presents a conceptually new way of strategy-making, as such practices have been employed throughout the whole of human history. However, it is important to emphasise that the history of war is full of examples in which new tactical and operational 'recipes' of old

CONCLUSIONS

strategy-making 'cocktail', skilfully adjusted to new social, political, economic and technological environments, have bestowed an enormous and immediate advantage on the first nation to employ them.

Since this is exactly what new-generation warfare and *gibridnaya voyna* attempt to do, the third important lesson to be drawn from this book concerns how the Western political and military leadership conceptualises, and responds, to the nature of the so-called Russian threat.

Lesson three: are the Russians coming?

To answer this question, it is important to understand the nature of the Kremlin's strategy and its actions on a tactical-operational level. The analysis in this book offers several important insights into the Kremlin's understanding of contemporary conflicts and the level of threat Russia poses to the West. First, as discussed in Chapter 7, Russian military thought draws a clear distinction between two different concepts: *gibridnaya voyna* and new-generation warfare. Whereas the former implies a mix of political, diplomatic, economic, information and other non-military means intended to subvert and undermine an adversary, the latter describes a full-scale military operation, preceded and accompanied by different non-military actions intended to weaken the adversary's military power and political resilience. An understanding of this division is important as it suggests that the Kremlin could apply different strategies in different scenarios, depending on whether it intends to engage in an open military confrontation.

Secondly, it is important to clarify the degree to which the theories of new-generation warfare and *gibridnaya voyna* represent Russian actions. As we have seen, some Western experts who have analysed the Russian discourse on these concepts rightly point to the fact that Russian military thinkers, scholars and experts were simply explaining their views of 'the operational environment and the nature of future war, and not proposing a new Russian way of warfare or military doctrine'.[46] Yet a close examination of the 2014 Russian Military Doctrine suggests that the ideas expressed by Gerasimov, Chekinov and Bogdanov regarding new-generation warfare and *gibridnaya voyna* are not simply their views alone. When discussing the nature of contemporary conflicts, for example, the doctrine echoes the theory of new-generation warfare, suggesting that such wars will involve 'a complex employment of military power, political, economic, informational and other non-military means'.[47] Moreover, when discussing Russian actions designed to prevent military confrontation, the doctrine clearly echoes the idea of *gibridnaya voyna*, stating that one of the goals of Russia's political and military leadership is 'a neutralisation of possible military threats and challenges by political, diplomatic and other non-military means'.[48] In other words, the concepts of new-generation warfare and *gibridnaya voyna* represent, at least to a certain degree, the Kremlin's strategic behaviour.

Thirdly, it is important to establish which of these two strategies Russia is employing against the West. While it is difficult to disagree with Galeotti that 'it would be naïve to consider today's Russia as a purely peaceable, defensive power',[49] it can be argued that the Kremlin is not actually seeking an open military confrontation with the West, nor does it believe that NATO presents a military threat per se. Russia's military doctrine does not define NATO as a military threat; instead, it defines the mere expansion of NATO—'bringing the military infrastructure of NATO member countries near the borders of the Russian Federation'—as a 'military risk' that has the potential to 'lead to a military threat under certain conditions'.[50]

However, as was discussed in Chapter 7, the Kremlin believes that the West is employing different non-military means and methods against Russia with the intention of undermining the Russian state, recreating the post-Cold War situation and ultimately bringing about the dissolution of Russia itself. It is reasonable to assume, therefore, that Russia has been employing the methods associated with *gibridnaya voyna* as a means of coercion and deterrence in an attempt to 'neutralise potential military risks and military threats through political, diplomatic and other non-military means'.[51]

There are two possible countermeasures that can be taken in response to this *gibridnaya voyna*: mirroring its methods or developing something different that better emphasises Western advantages and covers its weaknesses. However, before discussing these ways, it is important to focus on the fourth and most important point regarding the Russian threat—its novelty and significance. As already noted, there is nothing new in *gibridnaya voyna* as a strategy to achieve political goals without reverting to military force as such strategies have been employed throughout the whole of human history. The novelty of *gibridnaya voyna*, and thus its effectiveness, is instead represented by the tactical-operational adaptation of this well-known strategy to the realities of modern society and technology. It is important to emphasise that Russia was not the first to realise the political advantages of new communication platforms in the era of the internet, as well as the opportunities that the social freedoms of Western societies offer for those seeking to achieve certain

CONCLUSIONS

political gains—different social movements, terror organisations and even some states have engaged in similar practices long before Russia. The Kremlin, however, was the first to do so in a systematic, well-planned, well-executed and relatively well-funded way.

Military history teaches us that the best student of the previous war is the one who was defeated but did not reconcile themselves to defeat. Whether the Soviet Union was defeated by US political warfare, or whether it collapsed under the weight of its own deficiencies, is largely irrelevant in the domestic Russian context, as the current Russian political leadership, academia and military attributes its collapse to the actions of the West. It is unsurprising, therefore, that the theoretical and conceptual foundations of the place and role of non-military means and methods in inter-state confrontations have dominated the Russian political, military and academic discourse ever since the Cold War. As was also the case for inter-war Germany, which sought to identify the reasons for its defeat in the First World War, since the early 2000s Russia has been developing the theory of a non-military 'blitzkrieg' as a way to address its defeat in the Cold War (though *gibridnaya voyna*, by its very definition, is neither 'blitz' nor 'krieg', but a very long process of hostile inter-state relations).

It is important to point to the strengths and weaknesses of the Russian strategy of *gibridnaya voyna* in order to assess the threat that Russia actually poses to the West. On the hand, it seems that the Kremlin enjoys a temporary tactical-operational advantage, as while the West was swept up in the euphoria of victory in the Cold War and books on political warfare were gathering dust, the Russians were learning lessons from the Cold War and adjusting their means and methods to the social, political and technological environment of the twenty-first century. On the other, this tactical-operational advantage should not be overstated. The assumption that the advanced communication technologies of the twenty-first century, including the internet and social networks, make it easier to win information warfare and subvert political adversaries is similar to the assumption that other new pieces of technology, such as machine-guns or tanks, made it easier to achieve victory on land—an assumption that was frequently proven wrong. As Freedman puts it, 'the idea that an enemy can lob precision-guided thoughts into the collective mind of a population is far too simple', because 'the means by which humans receive and process information, as individuals or in groups, are complex and not so easy to manipulate'.[52]

The main weakness of *gibridnaya voyna*, however, is that it is not a new strategy, but rather a new tactical and operational way to execute an old and well-known strategy designed to achieve political goals; as with many other novel tactical-operational concepts, it offers an immediate, albeit temporary, advantage. Any conflict (whether military or not) is a duel, and due to its interactive nature, it is only a matter of time before tactical-operational gaps are bridged. The time that it will take the West to bridge these gaps depends on how quickly it is able to comprehend the conceptual and practical novelty of Russian *gibridnaya voyna* and how quickly it is able to adjust its own strategy of non-military confrontation to the social, political, economic and technological realities of the twenty-first century.

Lesson four: understanding Russia

In any discussion of the possibility of understanding Russia, it is difficult not to be reminded of the famous verse written by Fyodor Ivanovich Tyutchev in 1866:

> Russia cannot be known by the mind
> Nor measured by the common mile:
> Her status is unique, without kind—
> Russia can only be believed in.[53]

However, the idea that Russia 'is a riddle wrapped in a mystery inside an enigma', in Churchill's oft-cited words, is far from true; indeed, after making this statement, Churchill continued by saying 'but perhaps there is a key. That key is Russian national interest.'[54] As we saw with the analysis of Russia's doctrinal publications,[55] the key to which Churchill was referring is no less relevant for understanding Russia today than it was in 1939.

In the policy recommendations set out in his famous 'Long Telegram' that analysed the threat Soviet propaganda posed to the Western world, Kennan wrote: 'We must see that our public is educated to realities of Russian situation. I cannot over-emphasize the importance of this ... I am convinced that there would be far less hysterical anti-Sovietism in our country today if the realities of this situation were better understood by our people.'[56]

Although these words are as relevant today as they were in 1946, it would appear that the reason Russian *gibridnaya voyna* caught the West unprepared is that Kennan's recommendations were forgotten almost as soon as the Cold War ended.

It seems right to assume that before the dissolution of the USSR, the level of the West's understanding of Russia and vice versa were relatively similar—restricted to a limited number of academic and military experts—as the Cold

CONCLUSIONS

War limited the possibilities for people to travel (whether for academic purposes, work or just tourism). This situation has changed dramatically after the Cold War, yet the West has largely failed to take advantage of this: there is a vast difference between the numbers of Russians who have travelled or lived in the West and the numbers of Westerners who have travelled in the opposite direction. Whereas Russian academics view studying at Western universities, attending conferences in the West and publishing in Western journals as a normal (and sometimes even required) practice, the same is not necessarily true for Westerners, for whom spending time in Russia has remained a very limited experience, practised almost solely by those who specialise in Russian studies. Though there are many different reasons for this, the result is that Russia's current political leadership, academic community and citizens in general have a greater understanding of the West's strengths and weaknesses than the West does of Russia.

Furthermore, since the Cold War, the study of Russia in the West has declined significantly; as the US-based Association for Slavic, East European, and Eurasian Studies (ASEEES) stated in 2015: 'Russian studies within the social sciences are facing a crisis: an unmistakable decline in interest and numbers, in terms of both graduate students and faculty.'[57] There are three main reasons for this. The first is a general trend promoting quantitative methods in the social sciences, which have been gravitating away 'from emphasising expertise in particular regions toward more training in formal theory, sophisticated methodology, and comparative studies'.[58] The second is the reduction in government funding for Russia-related research, 'including cuts of over 50% to critical language training, and near complete elimination of advanced research fellowships for Americans on Russia and the region'.[59] ASEEES's report thus goes on to state:

> Due both to trends within political science away from area specific knowledge (and in the direction of broader theoretical and comparative studies and more sophisticated quantitative methods) and to a decline in interest on the part of the American public and government in Russia following the end of the Cold War, there are fewer faculty in political science departments who work on Russia than there were even a decade ago and also fewer PhD students. This is the gravest crisis facing the field.[60]

The third reason for the decline of Russian studies in the West was best described by Matthew Rojansk, Director of the Kennan Institute at the Wilson Center, who argued that the gaps in knowledge in Russian expertise created by a disregard of regional studies, in general, and lack of funding for

Russian studies, in particular, have been filled by experts from Central and Eastern Europe who 'have tended to maintain a much stronger capacity to understand and analyse Russia':[61] In other words, since Central and Eastern European experts have filled the ranks, the West has felt less need to educate its own experts to an extent that:

> The divergence of expertise between East and West had become so pronounced by the end of the last decade that in many intra-European and Euro-Atlantic forums, a de facto division of labour emerged in which representatives of Central and Eastern European member states assumed primary responsibility for analysing and developing collective policy recommendations towards Russia and the former Soviet space.[62]

Hence the general capacity of the West to understand Russia has been in decline since the end of the Cold War. As such, it is not surprising that the West was caught off-guard by Russia's actions in Crimea and eastern Ukraine. As Fiona Hill, director of the Brookings Institution's Centre on the United States and Europe, puts it 'they [Russians] do all these things, and sometimes they do signal quite clearly, but we missed a lot'.[63] As has been discussed throughout this book, the conceptual and political roots of *gibridnaya voyna* in the Russian academic, political and military discourse can easily be traced well back to the early 2000s. Whether there was somebody professional enough to interpret these signals, and powerful enough to change the course of Western policy, is a separate question that is beyond the scope of this book.

It is even less surprising that the Western discourse on Russian military and non-military activities was politicised so quickly, creating an unhelpful, alarmist perception of an almost existential threat to the Western world, rather than a calm, pragmatic and detailed understanding of Russian actions, their causes and, if required, ways to counteract them. Unfortunately, it seems that this situation is an outcome of Western incompetence in Russian affairs, together with the tendency of the Central and Eastern European states to exaggerate the threat posed by their eastern neighbour, unintentionally igniting an unhelpful and dangerous Cold War nostalgia. And, while the former has already been discussed, the latter rests at the core of the next lesson.

Lesson five: what next?

The conceptual relevance of Kennan's 'Long Telegram' to the nature of the contemporary confrontation between Russian and the West was already mentioned above, and therefore, it seems important to recall his previous advice:

CONCLUSIONS

Our first step must be to apprehend, and recognise for what it is, the nature of the movement with which we are dealing. We must study it with same courage, detachment, objectivity, and same determination not to be emotionally provoked or unseated by it, with which doctor studies an unruly and unreasonable individual.[64]

Consistent with Kennan's advice, this book has tried to separate the political and the conceptual discourse surrounding the ongoing confrontation between Russia and the West, and while the unhelpfulness of the former has already been discussed, it seems important to focus on the latter, trying to draw several useful recommendations for possible further actions. This book's analysis of the conceptual debate in the West and in Russia reveals that the current threat that Russia poses to the West is of a non-military nature. This does not mean that the military has no role to play as the demonstration of military force, or threat to use it, can magnify the effectiveness of non-military means (diplomatic, economic, international, etc.). Instead, it simply means that the Russian leadership prefers to operate in the non-military spectrum of confrontations, rather than in direct open military conflicts. Whether the Kremlin does so due to its unwillingness to resort to open conflict, or an inability to do, or simply because it believes that non-military means will be sufficient to achieve its political aims is less relevant because all are true to a certain degree. The most relevant part of this understanding is that the Kremlin employs non-military means and methods against the West, which, following the Russian understanding of this type of confrontation, are designed to destroy the political cohesion of the West from the inside. Hence by employing a carefully crafted hybrid of non-military means and methods that amplifies political, ideological, economic and other social divisions within Western society, the Kremlin is seeking to weaken the West's political integrity and resilience.[65]

Based on their experience of the Cold War, Russian analysts and military thinkers have focused on mastering the fundamentals of non-military confrontations, analysing historical examples, developing conceptual foundations, concocting an appropriate legislative framework intended to protect Russia from possible Western influence and building an effective infrastructure to execute offensive operations. In other words, by conceptualising *gibridnaya voyna* as something that the West wages against Russia, the Russians have mastered its methods by themselves. Moreover, if during the Soviet era, Russian non-military means and methods were used according to the Soviet ideology (so-called 'Active Measures'), contemporary Russia does it more pragmatically. While following the Marxist-Leninist ideology, the Soviet Union employed diplomatic, financial, economic, informational and other non-military means to weaken its political adversaries by amplifying the inter-class division;[66] contemporary Moscow aims to amplify any possible social, political, economic or ideological divisions that would help to undermine the political, economic and military cohesion and resilience of the West.

As already discussed, the Kremlin's actions do not represent a novel strategy-making concept, but rather a tactical and operational innovation, based on the ingenious implementation of new technologies and the exploitation of existing divisions within Western society. Therefore, in an attempt to successfully counteract Russian behaviour, the West, first and foremost, has to adapt its strategy of non-military confrontation to the social, political, economic and technological environment of the twenty-first century. In doing so, however, the leaders of Western states should seek to develop a strategy that uses the West's strengths and shields its weaknesses, rather than focusing on how to counteract Russian hacks, leaks, internet trolls or disinformation and propaganda operations. Thus, the key to the West's successes rests not in attempts to parry Russia's tactical-operational actions, but in a comprehensive strategy that will focus on the disease, not its symptoms.

The strategic idea of a non-military struggle that intends to subvert the enemy, rather than physically destroy it, is not foreign to the West, since the West successfully employed such a strategy during the Cold War.[67] Though the title of 'political war' has been misused just as much as the term 'war', the theory itself offers good conceptual foundations for the kind of strategic thinking that the West now requires. However, to be successful, the tactics associated with political warfare will need to be adapted to the realities of the twenty-first century. The analysis in this book offers several recommendations that may help with this process.

First, political warfare, like any other strategic concept, comprises two types of activities: offensive and defensive actions. Discussing the relation between these two types of actions in the context of the political warfare waged during the Cold War it is important to point to the fact that, though it included comprehensive offensive operations, its core was defensive.[68] Kennan had already identified the main weakness of the West as early as 1946, stating that the post-war European nations 'are tired and frightened by experiences of past', making them vulnerable to Soviet propaganda that sought to exploit existing social divisions: 'All persons with grievances, whether economic or racial, will be urged to seek redress not in mediation and compromise, but in defiant

CONCLUSIONS

violent struggle for destruction of other elements of society. Here poor will be set against rich, black against white, young against old, newcomers against established residents, etc.'[69]

To address this weakness, 'the Western defence of Europe' against Soviet propaganda took many forms, but first and foremost, it was directed at bridging the economic, political and social divisions that exposed the West to the Soviet propaganda machine, rather than trying to parry propaganda itself. This strategy was so successful that: 'What had been feared by European and American statesmen in 1946 no longer presented a serious threat in 1956. Europe had recovered, at least in the West, to a position of economic vitality, cultural self-esteem, and political stability.'[70]

This lesson of the past should serve as the basis for any attempt at adapting political warfare for the contemporary environment.

The second recommendation is based on one of Sun Tzu's observations concerning the nature of defence: 'he is skilful in defence whose opponent does not know what to attack'.[71] The West has many weaknesses that Russia is able to exploit. However, these need to be addressed at the strategic level rather than by seeking to defend against Russian tactical-operational actions by training police forces to better deal with 'genuine or manufactured protests', or 'blocking and undermining Russian propaganda campaigns' by installing 'proper controls on the flows of money from Russia' or increasing 'governance and legitimacy.'[72] All these are important, but they are all directed to counter Russian actions at a tactical-operational level rather than solve the strategic problem.

Mark Galeotti is correct to argue that 'the modern West—networked, globally-integrated, concerned with multiply real and perceived threats, and facing underlying crisis of confidence and legitimacy—has specific vulnerabilities the Russians are eagerly exploiting'. Yet this crisis of confidence and legitimacy also creates social, political, economic and ideological divisions that require a strategic solution, rather than tactical-operational attempts to respond to *gibridnaya voyna*. Although the West views the Baltic states and East European countries as being the most vulnerable to Russian non-military subversion,[73] the West needs to recognise that the main threat to these countries comes not from Russian disinformation operations but the lack of Western preparedness to defend these states.[74]

This leads to the book's third recommendation, that refers to the required nature of the West's defence. While Keir Giles is right in stating that 'during the Cold War, the strategy to defeat Soviet propaganda was built on two key principles: first, let an open society speak for itself; and second, expose the liar without worrying about every particular lie', it is difficult to agree with his conclusion that the applicability of this strategy is questionable and that the West's 'responses to information warfare need to mirror some of the technical approaches adopted by Russia'.[75] If there is something that can be learned from the experience of political warfare during the Cold War, it is that what makes Russia stronger weakens the West, and what makes the West stronger, weakens Russia. For example, as discussed in Chapter 7, while the politicisation of the West's *gibridnaya voyna* has led Russian society to close ranks against a common Western enemy, Russian interference in the 2016 US presidential elections has only furthered pre-existing political divisions.[76] The Cold War showed that the West's main strength is its freedom, and while James Sherr might be right in arguing that 'a free media should not be defenceless in the face of trolling, state-sponsored manipulation and cyber-attack'; by calling for 'non-defence arms of government (judicial, financial, regulatory)'[77] to protect it, it seems that he forgets that the main defence of a free media is the fact that it is free.

The fourth recommendation is based on the contradictory natures of the previous two, as freedom and openness expose social divisions to possible intensification, thus dwindling the capacity to protect them, and yet restricting freedom and openness harms the West's very forte that rests at the core of this protection. While there is no clear-cut formula for the winning balance between these two, based on the analysis of this book, it seems possible to point to two interconnected aspects that the West's leadership should take into account. Firstly, as already stated, the West has to shake the dust off the theory of political warfare, revisiting not only what was done, but also how. Moreover, in its study of that experience, the West should pay special attention to its defensive characteristics, rather than pondering ways to undermine the Soviet Union, not only because "the defensive form of warfare is intrinsically stronger than the offensive,"[78] but also because contemporary Russia enjoys a temporary tactical-operational advantage in both defending its weaknesses and exploiting the weaknesses of the West. Building its non-military defences the West has to remember that it is important to "approach calmly and with good heart [the] problem of how to deal with Russia,"[79] because political scare and alarmist behaviour are counterproductive, as they play into Russian hands, exacerbating social divisions within the West and threatening to diminish the West's main strength—its openness and freedoms.

The West should also consider trying to find a new balance between its strengths and weaknesses in its reliance on its Eastern European partners in interpreting Russia. As discussed above, during the last two decades the Western understanding of Russian affairs has been highly influenced by the opinions of scholars from the Central and Eastern European states. Though Rojansky is probably right to claim that these states have 'a much stronger capacity to understand and analyse Russia',[80] it is important to take into consideration the historical-cultural biases that many of these experts bring with them onto the West's decision-making table. While their insights on many aspects of Russian affairs may be invaluable, their advice on how to wage political warfare should be taken with extra caution. First, during the last round of this struggle (i.e., the Cold War) these countries were on the wrong side of the border, and while they might enjoy the outcome of the West's victory, they know and understand very little on how it was achieved. And secondly, since they were under the Soviet umbrella for the majority of the twentieth century, all are, in fact, quite young democracies, and appreciating liberal democratic values (which they definitely do) does not necessarily imply the wisdom required to protect them. It is not surprising that the politicisation of Russian hybrid warfare after 2014 was initiated by these very same Central and Eastern European states (that admittedly have good historic reasons to fear Russia more than any other states); but their suggestion to fight the Russian 'stick' with a Western 'stick' is not necessarily the best strategy. This does not imply, however, that the more experienced West has to silence its Eastern partners, as this is not the Western way of doing things, and, therefore, such an action would only undermine the West, rather than help it. This simply means that the more experienced West has to ground its actions on its own experience and expertise, and not be dragged by the alarmist agenda based on the unhelpful memory of the Cold War.

This leads to the final recommendation of this book. Although the book's occasional references to Kennan might suggest that the book is proposing a return to the practices of the Cold War, the opposite is true. Yes, Kennan's suggestions are important, as they were formulated at a time when Soviet propaganda enjoyed a tactical-operational advantage over the less-prepared West—something that is similar to the current situation. Yet it is also important to remember that it was Kennan's strategy of containment that led the West into the Cold War; and while it was, probably, the only solution at that time, it does not mean that it is so now.

The Cold War was not only a nuclear stand-off but also a period in Western history that was enormously expensive, both financially and politically.[81] Since a new Cold War would prove even more costly,[82] the West needs to learn its lessons and adjust Kennan's advice to the reality of the twenty-first century, rather than mimic them—Russia, after all, is not the Soviet Union; Europe is not devastated by war; and the United States has neither the resources, nor the political will, for a second Marshall Plan.

In order to avoid the repetition of another Cold War, the leaders of Western nations will need to make a series of difficult decisions in their relations with Russia. But since the West (in its political, rather than geographical sense) represents the most stable, experienced, economically powerful and progressive faction on this planet, it seems about time that it should start behaving as such. Not only because the alternative is far worse, but also because it might be that working together with Russia is more beneficial than fighting it. In the twenty-first century, the world faces dangers that are greater than any of those it has had to face since the end of the Second World War—from international terrorism to climate change—and another Cold War would simply be a lose–lose situation for the West and Russia, and for the rest of the world.

This outcome is not unavoidable. However, the leaders of Western nations need to recognise that they have made mistakes in their policy towards Russia. As Michael McFaul, former US ambassador to Russia and senior adviser to President Obama on Russian and Eurasian affairs, puts it:

> The mistake that was made 20 years ago was assuming Russia's a weak power, a declining power. Whether they're a great power or a middling power, we can argue about. But they are a major power, in the top 5 or 10 economies in the world, a top nuclear country in the world and now, given the investment Putin's made in the military, they're one of the major military powers in the world. Those trends are not changing in the next 20 or 30 years.[83]

Hence the West needs to recognise that Russia is a major power, that is going to stay around, alive and kicking, with Putin or without him, protecting its interests and unwilling to dance to the West's tune. And, therefore, there is little help in assessing that Moscow is 'politically isolated, economically sanctioned and with few options to improve its lot' or in how vulnerable 'an over-geared, under-invested, over-securitised and under-legitimate Russia may be',[84] not only because it does not represent the trend, but also because it misleads and creates an unhelpful delusion regarding the state of Russian affairs.

The Russian political system is different from the prevailing system in Western countries; of that there is no doubt. But this does not necessarily mean that it is weaker. During the twentieth century, the Russian people proved twice—in 1917 and again in 1991—that when they are unhappy with their leadership, they are willing and able to change it, regardless of the con-

CONCLUSIONS

sequences that may follow. However, this is unlikely to happen in the immediate future because the memory of the 1990s is still too fresh in the hearts and minds of the Russian people, and the economic and political stability created by the current regime is not going to be easily dismissed for an idea of material rights and freedoms (something that the Russians struggle to believe in anyway).[85] In other words, the West will need to accept not only that the Russian regime is Russia's business but also that the regime is likely to survive and that it would be beneficial to work with it rather than against it.

This leads to the second difficult decision that the West will need to take if it wants to avoid a new Cold War. In his recommendations to the West following the Ukrainian Crisis, James Sherr of Chatham House argues that: 'It is not our place to "promote democracy" in Russia or elsewhere, let alone regime change. Russian democracy is Russia's business. But Ukrainian democracy is not Russia's business, and neither is Ukraine's choice of partners.'[86]

Though undoubtedly true, Sherr forgets to mention the other side of the coin—Ukrainian democracy should not be the West's business either. In other words, as a direct continuation of the previous understanding that a strong Russia is going to continue to exist as a major political actor in the region, the West will need to accept Russia's role in that region. In the words of John J. Mearsheimer: 'Washington may not like Moscow's position, but it should understand the logic behind it. This is Geopolitics 101: great powers are always sensitive to potential threats near their home territory.'[87]

The West faces a choice between taking a politically difficult but responsible path, which would allow it to repair its relationship with Moscow, creating a situation where all sides win, and a more impulsive and reckless path, which may be politically easier, but in which everyone would lose.

Unfortunately, as this book's analysis of the Western discourse around Russian actions in Crimea and eastern Ukraine shows, it seems that the West has chosen to follow the latter path rather than the former. Whether this period will come to be known as a second Cold War or a hybrid war or any other voguish name is less important, ultimately, than the fact that we will be forced to bear its consequences.

P.S. A LESSON FOR RUSSIA

The West will clearly have to make a number of difficult decisions if it wants to avoid a new Cold War. However, this book's analysis of contemporary Russian political discourse also suggests that Russia will also need to make some difficult decisions if it wants to avoid the same scenario. The Russians, it seems, are talking themselves into a new round of unhelpful confrontations to the same extent and with the same speed as is the West.

Like the West, the Russian leadership should recognise that a return to the Cold War in the twenty-first century would be politically and economically devastating. As such, the recommendations in the previous chapter regarding making politically responsible decisions are just as applicable to Russia as they are to the West. Since the Kremlin believes that its foreign policy 'reflects the unique century-old role of Russia as a balancing factor in international affairs and the development of world civilisation',[1] Moscow needs to start balancing its relations with the West; in the process, it may find that certain political compromises can be made without any concessions whatsoever.

The analysis of Russian discourse on the nature of contemporary conflicts presented in this book has shown that the Russians are good students of Clausewitz, and they would do well to remember his lesson on the culminating point of victory: 'One must know the point to which it [war] can be carried in order not to overshoot the target; otherwise instead of gaining new advantages, one will disgrace oneself.'[2]

It seems that in its spat with the West, Russia has certain advantages—it comes more prepared and it knows the West's weaknesses much better than the West knows Russia's. Yet the fact that many Russians have travelled, studied and lived in the West is also a weakness. It does not necessarily mean that the Russians think that the West is better; it means that they know that it is different, and, when times are difficult, things that are different appear better than what one has. Moreover, the Kremlin should remember not only that this advantage is temporary but also that the West is a quick learner. While Clausewitz says that 'sometimes, stunned and panic-stricken, the enemy may lay down his arms, [and] at other times he may be seized by a fit of enthusiasm: there is a general rush to arm, and resistance is much stronger',[3] history teaches that, in dealing with the West, Russia may have more chances to find the latter type of enemy, rather than the former.

Since the Cold War, the West has made many mistakes—some were quickly recognised, others took years to understand, for some the West has been punished, and for others it punishes itself. The West has undoubtedly made mistakes in its approach toward Russia, too. Yet the Kremlin needs to remember that 'the superiority one has or gains in war is only the means and not the end',[4] and that punishing the West 'for assuming Russia's a weak power, a declining power',[5] is not the goal, but merely a way to teach the West a lesson (and it seems that the lesson has already been learned).

Taking these difficult decisions will not be easy for Russia or the West, and any improvement in their current relations will without doubt be a long and painful process. But the alternative is far worse. Both sides need to recognise that the world needs to change on the basis of mutual respect and understanding, as well as an ability to take responsibility for previous mistakes and to be ready to compromise. It is only by doing so that we will be able to avoid an uncertain future in which we are all bound to lose.

NOTES

INTRODUCTION

1. Lasconjarias, Guillaume and Jeffrey Larsen (eds), *NATO's Response to Hybrid Threats*, Rome: NATO Defence College, 2015.
2. Tsygankov, Pavel (ed.), *'Gibridnyye Voyny' v khaotiziruyushchemsya mire XXI veka* ['Hybrid wars' in the chaotic world of the twenty-first century], Moscow: Moscow University Press, 2015.
3. 'Inside the KGB: An Interview with Retired KGB Maj. Gen. Oleg Kalugin', CNN, https://web.archive.org/web/20070206020316/http://www.cnn.com/SPECIALS/cold.war/episodes/21/interviews/kalugin/ (accessed: 1 July 2017).
4. For example: Andrew, Christopher and Vasili Mitrokhin, *The Mitrokhin Archive: The KGB in Europe and the West*, London: Penguin Books, 2000; Lunev, Stanislav, *Through the Eyes of the Enemy: The Autobiography of Stanislav Lunev*, Washington, DC: Regnery Publishing, 1998; Earley, Pete, *Comrade J: The Untold Secrets of Russia's Master Spy in America After the End of the Cold War*, London: Penguin Books, 2007.
5. Von Clausewitz, Carl, *On War*, ed. and trans. Michael Howard and Peter Paret, Oxford: Oxford University Press, 2008, p. 13.

1. THE CONCEPTUAL FOUNDATIONS OF 'HYBRID WARFARE'

1. Hoffman, Frank, *Conflict in the 21st Century: The Rise of Hybrid Warfare*, Arlington: Potomac Institute for Policy Studies, 2007, p. 14.
2. Ibid.
3. For example: Nemeth, William, 'Future War and Chechnya: A Case for Hybrid Warfare', PhD diss., Monterey, Naval Postgraduate School, 2002; Morelock, Jerry, 'Washington as Strategist: Compound Warfare in the American Revolution, 1775–1783', in Huber, Thomas (ed.), *Compound Warfare: That Fatal Knot*, Fort Leavenworth: Combat Studies Institute, US Army Command and General Staff College, 2002, p. 78.

4. Qiao, Liang and Xiangsui Wang, *Unrestricted Warfare: China's Master Plan to Destroy America*, Panama City: Pan American Publishing Company, 2002.
5. Scobell, Andrew, 'Introduction to Review Essays on "Unrestricted Warfare"', *Small Wars and Insurgencies*, 11, 1 (Spring 2000), pp. 112–13; Cheng, Dean, '*Unrestricted Warfare*: Review Essay II', *Small Wars and Insurgencies*, 11, 1 (Spring 2000), pp. 122–9.
6. Thomas Moorer cited on the back cover of Qiao and Wang, *Unrestricted Warfare*.
7. Bunker, Robert, '*Unrestricted Warfare*: Review Essay II', *Small Wars and Insurgencies*, 11, 1 (Spring 2000), pp. 114.
8. For example: Luman, Ronald (ed.), *Unrestricted Warfare Symposium 2006: Proceedings on Strategy, Analysis, and Technology*, Laurel, MD: Johns Hopkins University Applied Physics Laboratory, 2006; Luman, (ed.), *Unrestricted Warfare Symposium 2008: Proceedings on Combating the Unrestricted Warfare Threat; Integrating Strategy, Analysis, and Technology*, Laurel, MD: Johns Hopkins University Applied Physics Laboratory, 2008.
9. Qiao and Wang, *Unrestricted Warfare*, p. 155.
10. Ibid.
11. Hoffman, *Conflict in the 21st Century*, p. 22.
12. Qiao and Wang, *Unrestricted Warfare*, p. xxi.
13. Ibid., p. 115.
14. Ibid., p. 117.
15. Ibid., pp. 115–23.
16. Ibid., pp. 129, 133, 129–48.
17. Ibid., pp. 146–7.
18. Ibid., p. 118.
19. Ibid., p. xx.
20. Ibid., see Chapter 7.
21. Ibid., p. 10.
22. Ibid., p. 36.
23. Ibid., pp. 36–43, 123–5.
24. Ibid., pp. 154, 155, 158.
25. Ibid., p. 162.
26. Ibid., pp. 161–4.
27. Ibid., pp. 164–8.
28. Ibid., p. 171.
29. Bunker, '*Unrestricted Warfare*: Review Essay II'; Van Messel, John, 'Unrestricted Warfare: A Chinese Doctrine for Future Warfare?', Master's thesis, Marine Corps University, Quantico, 2005.
30. Qiao and Wang, *Unrestricted Warfare*, p. 48.
31. Ibid., p. 61.
32. Ibid., pp. 155–61, 163.

33. Ibid., Chapter 3.
34. Ibid., p. xxi.
35. Ibid., p. 177.
36. Ibid., p. 178.
37. Ibid., pp. 179, 180.
38. Ibid., p. 182.
39. Ibid., pp. 182, 183.
40. Ibid., p. 184.
41. Ibid., p. 185.
42. Ibid., p. 155.
43. Ibid., p. 155; see also pp. 115–19.
44. Ibid., pp. xxi–xxii.
45. Lind, William, et al., 'The Changing Face of War: Into the Fourth Generation', *Marine Corps Gazette* (October 1989), pp. 22–6.
46. Van Creveld, Martin, *On Future War*, London: Brasseys, 1991.
47. Huntington, Samuel, *The Clash of Civilisations and the Remaking of World Order*, London: Simon and Schuster, 1997.
48. For example: Terriff, Terry, Aaron Karp and Regina Karp, (eds), *Global Insurgency and the Future of Armed Conflict*, New York: Routledge Press, 2007; Hammes, Thomas, *The Sling and the Stone: On War in the 21st Century*, St. Paul, MN: Zenith Press, 2004; Benbow, Tim, 'Talking 'Bout Our Generation? Assessing the Concept of "Fourth Generation Warfare"', *Comparative Strategy*, 27, 2 (2008), pp. 148–63.
49. Echevarria, Antulio, *Fourth Generation War and Other Myths*, Carlisle, PA: Strategic Studies Institute, 2005.
50. Lind, William, 'Understanding Fourth Generation War', *Military Review* (September–October 2004), p. 12.
51. Echevarria, *Fourth Generation War and Other Myths*, p. v.
52. Hammes, *Sling and the Stone*, p. 16.
53. Rogers, Clifford (ed.), *The Military Revolution Debate: Readings on the Military Transformation of Early Modern Europe*, Oxford: Westview Press, 1995; Parker, Geoffrey, *The Military Revolution: Military Innovation and the Rise of the West, 1500–1800*, Cambridge: Cambridge University Press, 1988; Boot, Max, *War Made New: Technology, Warfare and the Course of History, 1500 to Today*, New York: Gotham Books, 2006; Murray, Williamson and Macgregor Knox (eds), *The Dynamics of Military Revolution 1300–2050*, Cambridge: Cambridge University Press, 2001.
54. Hammes, *Sling and the Stone*, pp. 17, 18.
55. For example: Rogers, *Military Revolution Debate*; Parker, *Military Revolution*.
56. Lind, 'Understanding Fourth Generation War', p. 12.
57. Ibid.
58. Hammes, *Sling and the Stone*, pp. 18–20.

59. Ibid., p. 22.
60. Lind et al., 'Changing Face of War', pp. 12, 23.
61. Lind, 'Understanding Fourth Generation War', p. 12.
62. Lind et al., 'Changing Face of War', p. 23; also see Hammes, *Sling and the Stone*, pp. 22–31.
63. Lind, 'Understanding Fourth Generation War', p. 13.
64. Hammes, *Sling and the Stone*, pp. 23–30.
65. Lind, 'Understanding Fourth Generation War', pp. 12–13.
66. Lind et al., 'Changing Face of War', p. 23.
67. Hammes, *Sling and the Stone*, p. 31.
68. Hammes, Thomas, 'War Evolves into the Fourth Generation', *Contemporary Security Policy*, 26, 2 (2005), p. 197.
69. Ibid., pp. 197, 205.
70. Lind et al., 'Changing Face of War', p. 23.
71. Ibid., p. 24.
72. Hammes, 'War Evolves into the Fourth Generation', p. 206.
73. Hammes, *Sling and the Stone*, p. 208.
74. Echevarria, *Fourth Generation War and Other Myths*, p. 16.
75. Ibid.
76. Huber, Thomas, 'Napoleon in Spain and Naples: Fortified Compound Warfare', in *C610: The Evolution of Modern Warfare, Term I Syllabus/Book of Readings*, Fort Leavenworth: Combat Studies Institute, US Army Command and General Staff College, 1997.
77. Huber, *Compound Warfare*.
78. Huber, Thomas, 'Compound Warfare: A Conceptual Framework', in Huber, *Compound Warfare*, p. 1.
79. Roberts, Michael, 'The Military Revolution, 1560–1660', in Rogers, *Military Revolution Debate*.
80. Huber, 'Compound Warfare', pp. 1–2.
81. Ibid., p. 1.
82. Ibid., p. 5.
83. Ibid.
84. Ibid.
85. Ibid., p. 6.
86. Ibid., p. 7.
87. Ibid., pp. 3–4.
88. Ibid., p. 4.
89. See Morelock, Jerry, 'Washington as Strategist: Compound Warfare in the American Revolution' in Huber, *Compound Warfare*; Baumann, Robert, 'Compound War Case Study: The Soviets in Afghanistan', in Huber, *Compound Warfare*.

90. Huber, 'Compound Warfare', p. 5.
91. Hoffman, *Conflict in the 21st Century*, pp. 25–6.
92. Rumsfeld, Donald, 'The National Defense Strategy of the United States of America', Washington, DC, March 2005, p. v.
93. Ibid., p. i.
94. Hoffman, *Conflict in the 21st Century*, p. 26.
95. Rumsfeld, 'National Defense Strategy', p. 3.
96. Ibid.
97. Ibid., pp. 3–4.
98. Ibid., p. 3.

2. THE BIRTH OF 'HYBRID WARFARE'

1. Hoffman, Frank, '"Hybrid Threats": Neither Omnipotent nor Unbeatable', *Orbis*, 54, 3 (2010), p. 443.
2. Hoffman, *Conflict in the 21st Century: The Rise of Hybrid Warfare*, Arlington: Potomac Institute for Policy Studies, 2007, p. 27.
3. Hoffman, Frank, 'Hybrid Warfare and Challenges', *Joint Force Quarterly*, 52 (2009), pp. 35–6; Hoffman, *Conflict in the 21st Century*, pp. 26–8. Also see: Evans, Michael, 'From Kadesh to Kandahar: Military Theory and the Future of War', *Naval War College Review*, 56, 3 (2003), pp. 132–51; Blank, Stephen, 'The War That Dare Not Speak Its Name', *Journal of International Security Affairs*, 8 (2005); Gray, Colin, *Another Bloody Century: Future Warfare*, London: Weidenfeld & Nicolson, 2006; Arquilla, John, 'The End of War As We Knew It', *Third World Quarterly*, 28, 2 (2007), pp. 369–86; Hoffman, Bruce, '"The Cult of the Insurgent": Its Tactical and Strategic Implications', *Australian Journal of International Relations*, 61, 3 (2007), pp. 312–29; Robb, John, *Brave New War: The Next Stage of Terrorism and the End of Globalisation*, Hoboken: Wiley, 2007.
4. Hoffman, *Conflict in the 21st Century*, p. 30.
5. Hoffman, Frank, 'Hybrid Threats: Reconceptualising the Evolving Character of Modern Conflict', *Strategic Forum*, 240 (April 2009).
6. Hoffman, *Conflict in the 21st Century*, p. 29.
7. Hoffman, '"Hybrid Threats": Neither Omnipotent', p. 443.
8. Hoffman, 'Hybrid Threats: Reconceptualising', p. 5.
9. Hoffman, *Conflict in the 21st Century*, p. 14.
10. Ibid., p. 26.
11. Ibid., p. 29.
12. Ibid., pp. 29–30.
13. Hoffman, '"Hybrid Threats": Neither Omnipotent', p. 444.
14. Hoffman, 'Hybrid Threats: Reconceptualising', p. 5.
15. Hoffman, '"Hybrid Threats": Neither Omnipotent', p. 443.
16. Hoffman, 'Hybrid Warfare and Challenges', p. 35.

17. Hoffman, '"Hybrid Threats": Neither Omnipotent', p. 443.
18. Hoffman, 'Hybrid Threats: Reconceptualising', p. 6.
19. Hoffman, *Conflict in the 21st Century*, pp. 43–54.
20. Hoffman, '"Hybrid Threats": Neither Omnipotent', p. 455.
21. Hoffman, *Conflict in the 21st Century*, p. 35.
22. Ibid., p. 36; see also Hoffman, 'Hybrid Warfare and Challenges', p. 37; Hoffman, 'Preparing for Hybrid Wars', *Marine Corps Gazette*, 91, 3 (2007), pp. 57–61.
23. Hoffman, *Conflict in the 21st Century*, pp. 36–7; see also Hoffman, 'Hybrid Warfare and Challenges', p. 37; Hoffman, 'Preparing for Hybrid Wars'.
24. Hoffman, *Conflict in the 21st Century*, pp. 38–9.
25. Ibid., p. 40.
26. Hoffman, '"Hybrid Threats": Neither Omnipotent'.
27. Hoffman, *Conflict in the 21st Century*, pp. 61–2.
28. Hoffman, Frank and James Mattis, 'Future Warfare: The Rise of Hybrid Warfare', *Naval Institute Proceedings*, 132 (November 2005); Hoffman, Frank, 'How the Marines Are Preparing for Hybrid Wars', *Armed Forces Journal International*, 143, 8 (April 2006), p. 24; Hoffman, 'Hezbollah and Hybrid Wars: U.S. Should Take Hard Lesson from Lebanon', *Defense News*, 14 August 2006, p. 52; Hoffman, 'Further Thoughts on Hybrid Threats', *Small Wars Journal* (3 March 2009); Hoffman, 'Preparing for Hybrid Wars'; Hoffman, 'Hybrid Warfare and Challenges'; Hoffman, 'Hybrid Threats: Reconceptualising'; Hoffman, '"Hybrid Threats": Neither Omnipotent'.
29. Hoffman, *Conflict in the 21st Century*, p. 5.
30. For example: Carron, Ryan, *Hezbollah: Operational Art in Fourth Generation Warfare*, Newport: Naval War College, 2006; Cordesman, Anthony, 'Preliminary "Lessons" of the Israeli–Hezbollah War', working draft for outside comment, 2nd edn, Washington, DC: Center for Strategic and International Studies, 2006; Kreps, Sarah, 'The 2006 Lebanon War: Lessons Learned', *Parameters*, 37, 1 (2007), pp. 72–84; Biddle, Stephen and Jeffrey Friedman, *The 2006 Lebanon Campaign and the Future of Warfare: Implications for Army and Defense Policy*, Carlisle: Strategic Studies Institute, 2008; Farquhar, Scott (ed.), *Back to Basics: A Study of the Second Lebanon War and Operation CAST LEAD*, Fort Leavenworth: Combat Studies Institute Press, 2009; Johnson, David, *Hard Fighting: Israel in Lebanon and Gaza*, Santa Monica: RAND Corporation, 2011.
31. Williamson, Steven, *From Fourth Generation Warfare to Hybrid War*, Carlisle Barracks: US Army War College, 2009.
32. Bond, Margaret, *Hybrid War: A New Paradigm for Stability Operations in Failing States*, Carlisle Barracks: US Army War College, 2007.
33. Lasica, Daniel, *Strategic Implications of Hybrid War: A Theory of Victory*, Fort Leavenworth: U.S. Army Command and General Staff College, 2009.
34. Jordan, Larry, 'Hybrid War: Is the U.S. Army Ready for the Face of 21st-Century

Warfare?', Master's thesis, US Army Command and General Staff College, Fort Leavenworth, 2009).
35. McWilliams, Sean, *Hybrid War beyond Lebanon: Lessons from the South African Campaign 1976–1989*, Fort Leavenworth: US Army Command and General Staff College, 2009.
36. Fleming, Brian, *The Hybrid Threat Concept: Contemporary War, Military Planning and the Advent of Unrestricted Operational Art*, Fort Leavenworth: US Army Command and General Staff College, 2011; Joint Irregular Warfare Center, 'Irregular Adversaries and Hybrid Threats, An Assessment: 2011', Norfolk, 2011.
37. Thompson, Brian, 'Rules of Engagement in Hybrid Warfare Integrated into Operational Design', Master's thesis, Air Command and Staff College, Maxwell Air Force Base, 2010.
38. Johnson, David, *Military Capabilities for Hybrid War: Insights from the Israeli Defence Forces in Lebanon and Gaza*, Santa Monica: RAND Corporation, 2010. This was a preliminary publication from a bigger research project that was published the following year: Johnson, *Hard Fighting: Israel in Lebanon and Gaza*.
39. US Government Accountability Office, 'Hybrid Warfare: Briefing to the Subcommittee on Terrorism, Unconventional Threats', Washington, DC, 2010.
40. Smith, Marty, *Airpower in Hybrid War: Ethical Implications for the Joint Force Commander*, Newport: Naval War College, 2014.
41. Burbridge, Dean, *Employing U.S. Information Operations against Hybrid Warfare Threats*, Carlisle Barracks: US Army War College, 2013.
42. McCulloh, Timothy and Richard Johnson, *Hybrid Warfare*, MacDill Air Force Base: Joint Special Operations University Press, 2013.
43. McCulloh, Timothy, 'The Inadequacy of Definition and the Utility of a Theory of Hybrid Conflict: Is the "Hybrid Threat" New?', in McCulloh and Johnson, *Hybrid Warfare*, p. 16.
44. Ibid., pp. 16, 21–2, 30.
45. Ibid., pp. 16, 22, 30.
46. Ibid., pp. 16, 22–3, 31–2.
47. Ibid., pp. 16–17.
48. Johnson, Richard, 'Operational Approaches to Hybrid Warfare', in McCulloh and Johnson, *Hybrid Warfare*, pp. 71–100.
49. Ibid., p. 102.
50. Ibid., pp. 95–8, 102–3.
51. Ibid., p. 103.
52. For example: Weigley, Russell, *The American Way of War*, Bloomington: Indiana University Press, 1973; Echevarria, Antulio, *Towards an American Way of War*, Carlisle: Strategic Studies Institute, 2004; Lewis, Adrian, *The American Culture of War*, New York: Routledge, 2007; Gray, Colin, 'National Style in Strategy: The American Experience', *International Security*, 6, 2 (1981), pp. 21–47; Mahnken,

Thomas, 'The American Way of War in the Twenty-First Century', *Review of International Affairs*, 2, 3 (2003), pp. 73–84; Murray, Williamson, 'The Future of American Military Culture: Does Military Culture Matter?', *Orbis*, 43, 1 (1999), pp. 27–42.

53. For example: Farrell, Theo, *The Norms of War: Cultural Beliefs and Modern Conflict*, London: Lynne Rienner Publishers, 2005; Gray, Colin, *Out of the Wilderness: Prime Time for Strategic Culture*, Fort Belvoir: Defence Threat Reduction Agency, 2006; Black, Jeremy, *War and the Cultural Turn*, Cambridge: Polity Press, 2012; Heuser, Beatrice, *Nuclear Mentalities? Strategies and Beliefs in Britain, France and the FRG*, London: MacMillan, 1998; Hanson, Victor, *Carnage and Culture*, New York: Random House, 2001; Lynn, John, *Battle: A History of Combat and Culture*, Boulder: Westview Press, 2003.

54. Lewis, *American Culture of War*, p. 21.

55. Weigley, *American Way of War*, p. xviii; see also: Lock-Pullan, Richard, 'U.S. Military Strategy, Strategic Culture, and the "War on Terror"', in Owens, John and John Dumbrell, (eds), *America's 'War on Terrorism': New Dimensions in U.S. Government and National Security*, Lanham: Lexington Books, 2008; Gray, Colin, *Irregular Enemies and the Essence of Strategy: Can the American Way of War Adapt?*, Carlisle: Strategic Studies Institute, 2006; Buley, Benjamin, *The New American Way of War: Military Culture and the Political Utility of Force*, New York: Routledge, 2008; Lewis, *American Culture of War*, Chapter 1.

56. For example: Gray, *Irregular Enemies*; Echevarria, *Towards an American Way of War*; Record, Jeffrey, *Beating Goliath*, Washington, DC: Potomac Books, 2009; Mahnken, Thomas, *Technology and the American Way of War since 1945*, New York: Columbia University Press, 2010; Murray, 'Future of American Military Culture'.

57. Von Clausewitz, Carl, *On War*, edited and translated by Michae Howard and Peter Paret, Oxford: Oxford University Press, 2008, p. 28.

58. Buley, *New American Way of War*, p. 2; see also: Cohen, Eliot, 'Kosovo and the New American Way of War', in Bacevich, Andrew and Eliot Cohen (eds), *War over Kosovo*, New York: Columbia University Press, 2001; Gray, *Irregular Enemies*; Lewis, *The American Culture of War*.

59. The Command and General Staff School, *Principles of Strategy for an Independent Corps or Army in a Theater of Operations*, Fort Leavenworth: Command and General Staff School Press, 1936, p. 19.

60. Gray, *Irregular Enemies*, p. 44.

61. Murray, 'Future of American Military Culture'.

62. Kane, Thomas, *Military Logistics and Strategic Performance*, London: Frank Cass, 2001; Gray, *Irregular Enemies*, pp. 45–7.

63. Weigley, *American Way of War*, p. xxi; see also Mahnken, Thomas, *United States Strategic Culture*, Washington, DC: Science Applications International Corp., 2006, pp. 10–12.

64. Lock-Pullan, Richard, *US Intervention Policy and Army Innovation: From Vietnam to Iraq*, London: Routledge, 2006, p. 13; Dima Adamsky, *The Culture of Military Innovation: The Impact of Cultural Factors on the Revolution in Military Affairs in Russia, the US, and Israel*, Stanford: Stanford University Press, 2010, p. 80.
65. Mahnken, *Technology and the American Way of War*, p. 5; see also: Adamsky, *Culture of Military Innovation*, pp. 85–87; Gray, *Irregular Enemies*, pp. 35–6; Lewis, *American Culture of War*, pp. 32–4.
66. Lewis, *American Culture of War*, p. 33.
67. Murray, Williamson and MacGregor Knox, 'Conclusion: The Future behind Us', in Murray, Williamson and MacGregor Knox (eds), *The Dynamics of Military Innovation, 1300–2050*, New York: Cambridge University Press, 2001, pp. 178–9; Adamsky, *Culture of Military Innovation*, p. 87.
68. Lewis, *American Culture of War*, p. 22.
69. Murray, Williamson and Peter Mansoor (eds), *Hybrid Warfare: Fighting Complex Opponents from the Ancient World to the Present*, Cambridge: Cambridge University Press, 2012; see also: McWilliams, *Hybrid War beyond Lebanon*; McCulloh and Johnson, *Hybrid Warfare*.
70. Hoffman, Frank, 'Hybrid vs. Compound Wars', *Armed Forces Journal* (October 2009), pp. 1–2; see also Hoffman, 'Hybrid Warfare and Challenges'.
71. Hoffman, 'Hybrid vs. Compound Wars', p. 2.
72. For example: Palazzo, Albert and Antony Trentini, 'Hybrid, Complex, Conventional, Fourth-Generation Counterinsurgency: It's Decision that Still Matters Most', *Australian Army Journal*, 2, 1 (2010), pp. 71–91; Magen, Amichai, 'Hybrid War and the "Guliverization" of Israel', *Israel Journal of Foreign Affairs*, 5, 1 (2011), pp. 59–72; Schroefl, Josef and Stuard Kaufman, 'Hybrid Actors, Tactical Variety: Rethinking Asymmetric and Hybrid War', *Studies in Conflict & Terrorism*, 37, 10 (2014), pp. 862–80.
73. For example: US Department of the Army, *FM 3–0 Operations*, Washington, DC, February 2008 with amendments from 22 February 2011, Part 1.5; US Department of the Army, *ADP 3–0 Unified Land Operations*, Washington, DC, October 2011, p. 4.
74. Hoffman, '"Hybrid Threats": Neither Omnipotent', p. 441.
75. US Joint Chiefs of Staff, *Joint Publication 1–02, Department of Defense Dictionary of Military and Associated Terms*, 8 November 2010 (as amended through 15 February 2016).

3. READING EVGENY MESSNER: THE THEORY OF 'SUBVERSION-WAR' (*MYATEZHEVOYNA*)

1. Vladimirov, Alexander, *Osnovy obshchey teorii voyny* [Fundamentals of the general theory of war], vol. 1, Moscow: Universitet Sinergiya, 2013, p. 97.
2. Messner, Evgeny, *Posluzhnoy Spisok* [Service record], Buenos Aires, 1972. First

published as Tereshchuk, A. and I. Petrov (eds), *Evgeny Eduardovich Messner: Sud'ba Russkogo Ofitsera* [Evgeny Eduardovich Messner: the fate of the Russian officer], Saint-Petersburg: Izdatel'stvo Sankt-Peterburgskogo Universiteta, 1997, p. 23.
3. In the Soviet era, the University was renamed, and it is now known now as the Odessa I.I. Mechnikov National University (literally, 'Odessa National University named after I. I. Mechnikov').
4. Domnin, Igor, 'Ot Pervoy mirovoy do "Tret'yey Vsemirnoy": Zhiznennyy put' General'nogo shtaba polkovnika E. E. Messnera' [From the First World War to the 'third worldwide war': the life-path of General Staff Colonel E.E. Messner], in Savinkin, Alexander (ed.), *Khochesh' mira, pobedi myatezhevoynu! Tvorcheskoye naslediye E. Messnera* [If you want peace, win subversion-war! The creative heritage of E. Messner], Moscow: Voyennyy Universitet, Russkiy Put', 2005, pp. 20–1.
5. Messner, *Posluzhnoy Spisok*, p. 23.
6. Ibid., p. 25.
7. Ibid., p. 23; see also Domnin, 'Ot Pervoy mirovoy do "Tret'yey Vsemirnoy"', p. 5.
8. Messner, *Posluzhnoy Spisok*, pp. 10, 23, 27–8.
9. Ibid., pp. 10, 28–30.
10. Domnin, 'Ot Pervoy mirovoy do "Tret'yey Vsemirnoy"', p. 23.
11. Ibid., p. 25.
12. Messner, Evgeny, 'Poslednyaya pobeda Imperatorskoy armii' [The last victory of the Imperial Army], *Nashi Vesti*, 253 (1967).
13. Messner, Evgeny, 'Vpechatleniya podporuchika E.E. Messnera' [Second Lieutenant E.E. Messner's impressions], *Russkoe Slovo*, 537 (1974), p. 3.
14. Domnin, 'Ot Pervoy mirovoy do "Tret'yey Vsemirnoy"', pp. 27–34.
15. Ibid., pp. 34–49.
16. Alexandrov, Kirill, *Armia Generala Vlasova, 1944–1945* [The Army of General Vlasov, 1944–1945], Moscow: Yauza Eskimo, 2006, p. 184.
17. Messner, Evgeny, *Lik sovremennoy voyny* [The face of contemporary war], Buenos Aires: South American Division of the Institute for the Study of the Problems of War and Peace named after Prof. General N.N. Golovin, 1959; Messner, Evgeny, *Myatezh: Imya Tret'yey Vsemirnoy* [Subversion: the name of the third worldwide war], Buenos Aires: South American Division of the Institute for the Study of the Problems of War and Peace named after Prof. General N.N. Golovin, 1960; Messner, Evgeny, *Vseminaya Myatezhevoyna* [The worldwide subversion-war], Buenos Aires: South American Division of the Institute for the Study of the Problems of War and Peace named after Prof. General N.N. Golovin, 1971.
18. Messner, *Vseminaya Myatezhevoyna*, pp. 161–2.
19. Messner, *Myatezh: Imya Tret'yey Vsemirnoy*, p. 13.
20. Messner, *Vseminaya Myatezhevoyna*, pp. 79–84.
21. Ibid., p. 11.
22. Savinkin, Alexander, 'Groznaya Opasnost' Vsemirnoy Myatezhevoyny' [The threat-

ening danger of the worldwide subversion-war], in Savinkin, *Khochesh' mira, pobedi myatezhevoynu*, p. 575

23. Gracheva, Tatiana, 'Postizheniye voyny: myatezh kak sostoyaniye dushi, tsel' i sposob zavoyevaniya' [Comprehending war: subversion as a state of soul, a purpose and a way of conquest], in Messner, Evgeny, *Vsemirnaya Myatezhvoyna* [Worldwide subversion-war], Moscow: Zhukovskoye Pole, 2004, p. 357.
24. Morozov, Evgeny, 'Predisloviye kommentatora' [Commentator's foreword], in Messner, *Vsemirnaya Myatezhvoyna*, p. 8.
25. Domnin, Igor, 'Kratkiy ocherk voyennoy mysli Russkogo zarubezh'ya' [A short outline of the military thought of Russian émigrés], in Savinkin, Alexander (ed.), *Voyennaya mysl' v izgnanii: tvorchestvo russkoy voyennoy emigratsii* [Military thought in exile: the oeuvre of Russian military émigrés], Moscow: Voyennyy Universitet, Russkiy Put', 1999, pp. 523–4.
26. Maneev, Igor, 'Voyenno–filosofskiye idei Russkogo zarubezh'ya pervoy poloviny XX veka' [The military–philosophical ideas of Russian émigrés in the first part of the twentieth century], PhD thesis, Military University of the Ministry of Defence of the Russian Federation, 2011; Domnin, 'Kratkiy ocherk voyennoy mysli'.
27. Toffler, Alvin and Heidi Toffler, *War and Anti-war: Survival at the Dawn of the 21st Century*, Boston: Little, Brown and Company, 1993, pp. 33–85.
28. Boot, Max, *War Made New: Technology, Warfare and the Course of History, 1500 to Today*, New York: Gotham Books, 2006, pp. 13–15.
29. Murray, Williamson and Macgregor Knox, 'Thinking about Revolution in Warfare', in Murray, Williamson and Knox, *Dynamics of Military Revolution*, pp. 6–11.
30. Murray, Williamson, 'Thinking about Revolution in Military Affairs', *Joint Force Quarterly*, 16 (1997), p. 71.
31. Messner, *Myatezh: Imya Tret'yey Vsemirnoy*, p. 8.
32. Murray, 'Thinking about Revolution in Military Affairs', pp. 71–3.
33. Messner, *Myatezh: Imya Tret'yey Vsemirnoy*, p. 8.
34. Ibid., pp. 8–33.
35. Pinker, Steven, *The Better Angels of Our Nature: A History of Violence and Humanity*, London: Penguin Books, 2012.
36. Messner, *Myatezh: Imya Tret'yey Vsemirnoy*, p. 32.
37. Ibid., p. 5.
38. Ibid., p. 32.
39. Lind, William, et al., 'Changing Face of War: Into the Fourth Generation', *Marine Corps Gazette* (October 1989), pp. 25–6.
40. Messner, *Myatezh: Imya Tret'yey Vsemirnoy*, p. 58.
41. See Chapter 1.
42. Hammes, Thomas, *Sling and the Stone: On War in the 21st Century*, St. Paul, MN: Zenith Press, 2004, pp. 31, 208.
43. Messner, Evgeny, 'Gore i pobezhdennym i pobeditelyam' [Grief to the defeated and to victors], *Segodnya*, 71 (1931).

44. 'An officer of the ex-Russian Imperial Army' [Evgeny Messner], 'Die Dekadenz in der Kriegskunst', *Wissen und Wehr* (January 1926), p. 101 (in German).
45. Messner, *Lik sovremennoy voyny*, p. 11.
46. Messner, *Myatezh: Imya Tret'yey Vsemirnoy*, p. 43.
47. Ibid., pp. 26–8.
48. Messner, *Lik sovremennoy voyny*, p. 37.
49. Lind et al., 'Changing Face of War', p. 23.
50. Messner, *Lik sovremennoy voyny*, p. 4.
51. Ibid., p. 5.
52. Ibid., pp. 4–5.
53. Messner, *Lik sovremennoy voyny*, p. 5.
54. Messner, *Myatezh: Imya Tret'yey Vsemirnoy*, p. 43.
55. Ibid., p. 36.
56. Ibid., p. 95.
57. Messner, *Lik sovremennoy voyny*, p. 29.
58. Messner, *Myatezh: Imya Tret'yey Vsemirnoy*, p. 95; Messner, *Lik sovremennoy voyny*, p. 30.
59. Messner, *Myatezh: Imya Tret'yey Vsemirnoy*, p. 91.
60. Ibid., p. 96.
61. Ibid., p. 97.
62. Messner, *Lik sovremennoy voyny*, p. 29.
63. Messner, *Myatezh: Imya Tret'yey Vsemirnoy*, p. 97.
64. Ibid., pp. 97–8.
65. Ibid., p. 43.
66. Messner, *Lik sovremennoy voyny*, p. 7.
67. Lind et al., 'Changing Face of War', pp. 23, 26.
68. Messner, *Vseminaya Myatezhevoyna*, p. 12.
69. Messner, *Lik sovremennoy voyny*, p. 48.
70. Messner, *Myatezh: Imya Tret'yey Vsemitnoy*, pp. 88–9.
71. Messner, *Vseminaya Myatezhevoyna*, p. 10.
72. Ibid., p. 14.
73. Messner, *Lik sovremennoy voyny*, p. 13.
74. Ibid.
75. Ibid., p. 12.
76. Ibid., p. 13.
77. Ibid., pp. 13, 15.
78. Warren, Lewis, *King John*, London: Methuen, 1991.
79. Dull, Jonathan, *A Diplomatic History of the American Revolution*, New Haven: Yale University Press, 1985.
80. See Chapter 1.
81. For example: Cîrstea, Beng Sorin, 'Information Operations in The Context of

Hybrid Warfare', *Romanian Military Thinking*, 4 (2015), pp. 41–55; Sawa-Czajka, Elżbieta, 'Rebel-War in Ukraine', *Confrontation and Cooperation: 1000 Years of Polish–German-Russian Relations*, 1 (2014), pp. 25–31.
82. Messner, *Vseminaya Myatezhevoyna*, p. 8.
83. Ibid., p. 9.
84. Ibid., p. 64.
85. Ibid., p. 164.
86. Ibid., p. 10.
87. Messner, *Lik sovremennoy voyny*, p. 11.
88. Messner, *Vseminaya Myatezhevoyna*, p. 66.
89. Ibid., p. 78.
90. Messner, *Myatezh: Imya Tret'yey Vsemirnoy*, p. 5.
91. Ibid., pp. 90–2.
92. Messner, *Vseminaya Myatezhevoyna*, pp. 78–9.
93. Messner, *Myatezh: Imya Tret'yey Vsemirnoy*, p. 94.
94. Ibid.
95. Messner, *Vseminaya Myatezhevoyna*, p. 77.
96. Messner, *Myatezh: Imya Tret'yey Vsemirnoy*, p. 5.
97. Messner, Evgeny, 'V'yetnamskaya zagadka' [Vietnam's riddle], *Nashi Vesti*, 269 (1968).
98. Messner, *Vseminaya Myatezhevoyna*, pp. 79–85.
99. Messner, *Myatezh: Imya Tret'yey Vsemirnoy*, p. 93.
100. Messner, *Vseminaya Myatezhevoyna*, p. 26.
101. Ibid., p. 33.
102. Ibid., p. 45.
103. Ibid., p. 150.
104. Ibid., p. 93.
105. Messner, *Lik sovremennoy voyny*, p. 18.
106. For example: Domnin, Igor and Alexander Savinkin, 'Assimetrichnoe Voevanie' [Assymentric warfare], *Otechestvennyye Zapiski*, 5 (2005); Pavlushenko, Mikhail, Vladimir Zyzin and Yury Ol'hovnik, 'Myatezhevoyna kak forma tsivilizatsionnogo stolknoveniya "Zapad-Vostok"' [Subversion-war as a form of the civilisational clash 'West–East'], *Obozrevatel'-Observer*, 5 (2007), pp. 13–19; Biryukov, Sergey and Alexander Davydov, 'Konfliktnaya model' formirovaniya natsional'noy politicheskoy identichnosti: kontseptsiya "matezhevoyny" E. Messnera (na primere Bosnii)' [The conflict model of the formation of the national identity: the concept of subversion-way by E. Messner (case study: Bosnia)]', *Politicheskiye Instituty i Protsessy*, 2 (2014), pp. 132–45; Nesterov, Aleksey, 'Netraditsionnyye ugrozy voyennoy bezopasnosti Rossiyskoy Federatsii v ramkakh modernizatsii NATO' [Non-traditional threats to the security of the Russian Federation due to the modernisation of NATO]', *Vlast'*, 5 (2013), pp. 186–8; Morozov,

'Predisloviye kommentatora'; Savinkin, 'Groznaya Opasnost' Vsemirnoy Myatezhevoyny'.

107. Solov'ev, V. and A. Dremkov, 'Yeshche raz o predmete i strukture voyennoy nauki' [One more time on the subject and structure of military science], *Voyennaya Mysl'*, 9 (1994), pp. 34–5.

108. See: Savinkin, Alexander, (ed.), *Russkaya voyennaya doktrina* [Russian military doctrine], Moscow: Gumanitarnaya akademiya vooruzhennykh sil, 1994; Savinkin (ed.), *Russkoye zarubezh'ye: gosudarstvenno-patrioticheskaya i voyennaya mysl* [Russian émigrés: political-patriotic and military thought], Moscow: Gumanitarnaya akademiya vooruzhennykh sil, 1994; Savinkin (ed.), *K poznaniyu Rossii: Vzglyady russkikh mysliteley nachala XX veka* [Comprehending Russia: the views of the Russian thinkers in the beginning of the twentieth century], Moscow: Gumanitarnaya akademiya vooruzhennykh sil, 1994; Savinkin (ed.), *Kakaya armiya nuzhna Rossii? Vzglyad iz istorii* [What army does Russia need? A view from history], Moscow: Voyennyy Universitet, 1995; Savinkin (ed.), *Dusha armii: Russkaya voyennaya emigratsiya o moral'no-psikhologicheskikh osnovakh rossiyskoy vooruzhennoy sily* [The soul of army: Russian military émigré writing on the moral-psychological foundations of the Russian Armed Forces], Moscow: Voyennyy universitet, Nezavisimyy voyenno-nauchnyy tsentr 'Otechestvo i Voin', Russkiy put', 1997; Savinkin, *Voyennaya mysl' v izgnanii*; Savinkin (ed.), *Ofitserskiy korpus Russkoy Armii: Opyt samopoznaniya* [The officer corps of the Russian army: the experience of self-knowledge], Moscow: Voyennyy Universitet, Russkiy put', 2000; Savinkin, *Khochesh' mira, pobedi myatezhevoynu!*; Savinkin (ed.), *Groznoye oruzhiye: Malaya voyna, partizanstvo i drugiye vidy asimmetrichnogo voyevaniya v svete naslediya russkikh voyennykh mysliteley* [Formidable weapon: small war, partisans and other types of asymmetric warfare in light of Russian military thinkers' heritage], Moscow: Voyennyy Universitet, Russkiy put', 2007.

109. Savinkin, Alexander, 'Zashchita Rossii' [The Protection of Russia], in Savinkin, *Russkoye zarubezh'ye*, p. 8.

110. For example: Kamenev, Anatoliy, *Tragediya russkogo ofitserstva: Uroki istorii i sovremennost* [The tragedy of the Russian officer corps: the lessons of history and modernity], Moscow: Signal, 1999; Yershov, Vitaliy, *Rossiyskoye voyenno-politicheskoye zarubezh'ye v 1918–1945 gg.* [Russian military–political émigrés in 1918–45], Moscow: Moskovskii gosudarstvennii universitet servisa, 2000; Dyrin, Anatoliy, *Patrioticheskaya ideya i voyennaya doktrina dlya budushchey Rossii v literature russkogo zarubezh'ya pervoy poloviny XX veka* [The political ideas and military doctrine for future Russia in Russian émigré literature in the first part of twentieth century], Moscow: Megapir, 2008.

111. For example: Sotnikov, Sergey, 'Istoriko–kul'turnaya deyatel'nost' rossiyskoy voyennoy emigratsii vo Frantsii v 1920–1945 godakh' [The historical–cultural activ-

ity of Russian military émigrés in France in 1920–45], *Servis v Rossii i za rubezhom*, 8 (2012); Ustinkin, Sergey and Alexander Shaposhnikov, 'Vklad russkoy voyenno-politicheskoy emigratsii v nauku o voyne (20–30 gg. XX v.)' [The contribution of the Russian military–political émigrés to the science of war (1920–30)]', *Vestnik Akademii Voyennykh Nauk*, 3, 44 (2013), pp. 172–81; Muromtseva, Lyudmila, 'Voyenno-patrioticheskiye traditsii v zhizni rossiyskoy emigratsii' [The military–patriotic traditions in the life of Russian émigrés], in Koz'menko, V. and V. Kerov (eds), *Voyny i voyennyye konflikty v istorii Rossii: k 70-letiyu velikoy pobedy* [Wars and military conflict in the history of Russia: on the seventieth anniversary of the great victory], Moscow: Rossiyskiy Universitet Druzhby Narodov, 2015, pp. 251–62; Sakun, Sergey and Valery Kiselev, 'Rossiyskaya voyennaya klassika v sisteme moral'no-psikhologicheskogo obespecheniya voysk (sil)' [Russian military classics on the system of moral–psychological support of troops (forces)], *Voyennaya Mysl'*, 2 (2016), pp. 34–40.

112. Domnin, Igor, 'Voyennaya kul'tura russkogo zarubezh'ya' [The military culture of the Russian émigré], in Grigor'yev, Alexander (ed.), *Filosofiya Voyny* [The philosophy of war], Moscow: Ankil-Voin, Rossiyskiy Voyennyy Sbornik, 1995, pp. 8–9.

113. Konyshev, Valeriy and Alexander Sergunin, *Voyennaya strategiya sovremennogo gosudarstva* [The military strategy of the contemporary state], Saint-Petersburg: NPPL 'Rodnyye Prostory', 2012, p. 133.

114. Vladimirov, *Osnovy obshchey teorii voyny*, vol. 1, p. 48.

115. Gracheva, 'Postizheniye voyny', p. 341.

116. Babich, Vladimir, 'O novom podkhode k analizu sovremennogo protivoborstva i nekotorykh drugikh problemakh' [On the new approach to the analysis of the contemporary confrontation and some other problems], *Voyennaya Mysl'*, 3 (2008), p. 38.

117. Radikov, Ivan, 'Voyna v XXI veke i novaya semantika voyennoy doktriny Rossii' [War in the twenty-first century and the new semantics of Russia's military doctrines], *Istoricheskiye, filosofskiye, politicheskiye i yuridicheskiye nauki, kul'turologiya i iskustvovedeniye. Voprosy teorii i praktiki*, 5, 55 (2015), p. 151.

118. For example: Konopatov, Sergey and Vladimir Yudin, 'Traditsionnyy smysl ponyatiya voyna ustarel' [The traditional meaning of the term war is outdated], *Voyennaya Mysl'*, 1 (2001), pp. 53–7; Babich, 'O novom podkhode k analizu sovremennogo protivoborstva'; Radikov, 'Voyna v XXI veke'.

119. For example: Kuz'movich, Aleksey, 'Evolyutsiya vzglyadov na teoriyu sovremennoy voyny' [The evolution of views on the theory of contemporary war], *Armiya i obshchestvo*, 1, 33 (2013), pp. 138–41; Vilovatykh, Anna, 'Koye-chto o tekhnologiyakh global'nogo dominirovaniya na primere "Strategii nepriyatnykh deystviy"' [Commentary on the technologies of global dominance, the case of the 'strategies of indirect actions'], *Diskurs-Pi*, 4, 11 (2014), pp. 76–9; Afanas'yeva,

Yuliya, 'Massovaya kommunikatsiya kak nesilovoy faktor vedeniya voyny: istoriya voprosa i sovremennoye sostoyaniye' [Mass communications as a non-violent factor of waging war: the history of the subject and its contemporary state], *Armiya i obshchestvo*, 2, 32 (2012), pp. 65–70; Lyutkene, Galina, 'Nekotoryye novshestva v interpretatsii voyny na zapade' [Some novelty in the interpretation of war in the West], *Vestnik Voyennogo universiteta*, 2, 22 (2010), pp. 170–4; Kaftan, Vitaliy, 'Partizanskaya voyna i terrorizm: k proyasneniyu sushchnosti ponyatiy' [Partisan war and terrorism: the clarification of definitions], *Prostranstvo i Vremya*, 3, 9 (2012), pp. 88–93.
120. Messner, *Lik sovremennoy voyny*, p. 6.

4. NET-CENTRIC AND INFORMATION WARS: MODERN THEORIES OF SUBVERSION

1. Nekrasov, Stanislav, 'A.G. Dugin: Nastoyashchiy Postmodern' [A.G. Dugin: the real postmodern], *Diskurs-Pi*, 1, 1 (2001), pp. 43–52; Darczewska, Jolanta, 'The Anatomy of Russian Information Warfare: The Crimean Operation, A Case Study', *Point of View*, vol. 42, Warsaw: Centre for Eastern Studies, 2014, p. 14.
2. Dugin, Aleksandr, *Russkaya Veshch'* [Russian thing], Moscow: Arktogeya, 2001; Dugin, Filosofiya voyny [The philosophy of war], Moscow: Yauza, Eksmo, 2004; Dugin, *Geopolitika Postmoderna* [Postmodern geopolitics], Saint-Petersburg: Amfora, 2007; Dugin, *Sotsiologiya geopoliticheskikh protsessov Rossii* [The sociology of the geopolitical processes of Russia], Moscow: Lomonosov Moscow State University, 2010; Dugin, *Russkaya voyna* [Russian war], Moscow: Algoritm, 2015; Dugin, *Voyna kontinentov: sovremennyy mir v geopoliticheskoy sisteme koordinat* [A war of continents: the contemporary world in the geopolitical system of coordinates], Moscow: Akademicheskiy Proyekt, 2015.
3. Dugin, *Sotsiologiya geopoliticheskikh protsessov*, pp. 31–2.
4. Dugin, *Filosofiya voyny*, pp. 155–6.
5. Dugin, *Geopolitika Postmoderna*, p. 16.
6. Ibid., pp. 651–700.
7. For example: Dugin, Aleksandr, 'Teoreticheskiye osnovy setevykh voyn' [The theoretical grounds of network wars], *Informatsionnye voyny*, 1, 5 (2008), pp. 2–9; Dugin, 'Setetsentricheskiye voyny' [Net-centric wars], *Informatsionnye voyny*, 1, 5 (2008), pp. 10–16.
8. For example: Dugin, *Voyna kontinentov*, pp. 240–58.
9. Wogaman, Donald, *Network Centric Warfare: An Emerging Warfighting Capability*, Quantico, VA: Marine Corps Combat Development Command, 1998.
10. Cebrowski, Arthur and John Garstka, 'Network-Centric Warfare: Its Origin and Future', *Proceedings Magazine*, 124, 1, 1, 139 (January 1998), pp. 28–35.
11. Alberts, David, John Garstka and Frederick Stein, *Network Centric Warfare: Developing and Leveraging Information Superiority*, Washington, DC: Command

and Control Research Program, 1999; Alberts, David, et al., *Understanding Information Age Warfare*, Washington, DC: Command and Control Research Program, 2001.
12. Alberts, Garstka and Stein, *Network Centric Warfare*, p. 88.
13. Ibid., p. 91.
14. For example: Vladimirov, Alexander, *Osnovy obshchey teorii voyny* [Fundamentals of the general theory of war], vol. 1, Moscow: Universitet Sinergiya, 2013, pp. 405–44; Kiselov, V. and I. Ryabchenko, 'Novyye formy razvedki i ikh osobennosti: printsipy takticheskoy razvedki v usloviyakh setetsenricheskikh deystviy' [New forms of intelligence and their features: the principles of tactical intelligence under the conditions of net-centric warfare], *Armeyskiy sbornik*, 1, 235 (2014), pp. 36–40; Khamzatov, M., 'Adaptatsiya k sovremennosti: vliyaniye kontseptsii setetsentricheskoy voyny na kharakter boyevykh operatsiy' [The adaptation to modernity: the influence of the concept of net-centric war on the character of combat operations], *Armeyskiy sbornik*, 1, 235 (2014), pp. 41–3; Zinov'ev, Vyacheslav, Anatoly Koldinov and Nikolay Gruzdev, 'Perspektivy primeneniya informatzionnyh setey v voennom dele' [Prospects for the use of informational networks in military affairs], *Informatsionnye voyny*, 1, 33 (2015), pp. 37–40; Kondrat'yev, A., 'Kogda "setetsentrizm" pridyot v Rossiyskuyu armiyu?' [When will 'net-centrism' come to the Russian Army?], *Armeyskiy sbornik*, 5, 239 (2014), pp. 6–11; Dolgopolov, A., 'Yeshcho raz o setetsentricheskikh voynakh' [One more time on net-centric wars], *Armeyskiy sbornik*, 2, 248 (2015), pp. 3–6.
15. Dugin, 'Teoreticheskiye osnovy setevykh voyn', p. 2.
16. Dugin, *Voyna kontinentov*, p. 241.
17. Dugin, 'Setetsentricheskiye voyny', p. 10.
18. Dugin, 'Teoreticheskiye osnovy setevykh voyn', p. 4.
19. Ibid., p. 5.
20. Ibid.
21. Ibid.
22. Dugin, *Voyna kontinentov*, p. 242.
23. Dugin 'Teoreticheskiye osnovy setevykh voyn', pp. 5–6.
24. Ibid., p. 6.
25. Ibid.
26. Ibid., p. 7.
27. Ibid.
28. Ibid.
29. Ibid., pp. 7–8.
30. Ibid., p. 8.
31. Dugin, *Voyna kontinentov*, p. 242.
32. Ibid., p. 247.
33. Ibid., pp. 248–9.

34. Dugin, 'Teoreticheskiye osnovy setevykh voyn', p. 3.
35. Ibid., p. 8.
36. Ibid.
37. Ibid.
38. Ibid., p. 9.
39. Dugin, *Voyna kontinentov*, pp. 250–8.
40. Ibid., p. 250.
41. Dugin, 'Teoreticheskiye osnovy setevykh voyn', p. 9.
42. Dugin, *Voyna kontinentov*, p. 252.
43. For example: Savin, Leonid, *Setetsentrichnaya i setevaya voyna: Vvedeniye v kontseptsiyu* [Net-centric and network war: an introduction to the concept], Moscow: Yevraziyskoye dvizheniye, 2011; Savin, *Setetsentrichnyye boyevyye deystviya: Novyye sposoby vedeniya voyny; kak Amerika stroit imperiyu* [Net-centric combat operations: new methods of waging wars; how America builds an empire], Saint-Petersburg: Piter, 2016; Zariffulin, Pavel, 'Setevaya voyna na Severnom Kavkaze' [Network war in the north Caucasus], *Informatsionnye voyny*, 2, 6 (2008), pp. 37–41; Korovin, Valery, 'Setevaya voyna Ameriki protiv Rossii na primere Chechni' [The network war of America against Russia, the case of Chechnya], *Informatsionnye voyny*, 2, 6 (2008), pp. 42–6; Bovdunov, Alexander, 'Nepravitel'stvennyye organizatsii: setevaya voyna protiv Rossii' [Non-government organisations: network war against Russia], *Informatsionnye voyny*, 3, 7 (2008), pp. 30–9; Savin, Leonid, 'Ukraina v setevoy voyne' [Ukraine in network war], *Informatsionnye voyny*, 3, 7 (2008), pp. 42–51; Kanishchev, Pavel, 'Setevaya voyna SSHA protiv Rossii: pole boya: Gruziya' [The network war of the United States against Russia: the battlefield: Georgia], *Informatsionnye voyny*, 3, 7 (2008), pp. 52–6; Korovin, Valery, *Glavnaya voyennaya tayna SSHA. Setevyye voyny: 'tsvetnyye revolyutsii', taynyye zagovory i spetsoperatsii, podryvnaya propaganda, 'promyvaniye mozgov', voyny 21 veka* [The main military secret of the USA network wars: 'colour revolutions', secret plots and special operations, subversive propaganda, 'brainwashing', the wars of the twenty-first century], Moscow: Yauza, Eksmo, 2009; Filimonov, Georgy, *Kulturno–informatsionnie mekhanizmi vnewnei politiki SSHA: istoki i novaya real'nost'* [The cultural–informational mechanisms of US foreign policy: the origins and new reality], Moscow: People's Friendship University of Russia, 2012, pp. 189–206.
44. Panarin, Igor, *Informatsionnaya voyna i geopolitika* [Information war and geopolitics], Moscow: Pokolenie, 2006, p. 165.
45. For example: Panarin, Igor, *Pervaya mirovaya informatsionnaya voyna: razval SSSR* [The first world information war: the dissolution of the USSR], Saint-Petersburg: Piter, 2010; Panarin, *Informatsionnyye voyna i kommunikatsii* [Information war and communications], Moscow: Goryachaya Liniya-Telekom, 2015; Panarin, Igor and Lyubov' Panarina, *Informatsionnaya voyna i mir* [Information war and the world], Moscow: OLMA-PRESS, 2003.

46. Panarin and Panarina, *Informatsionnaya voyna i mir*, pp. 20–1.
47. Panarin, *Pervaya mirovaya informatsionnaya voyna*, p. 24.
48. Grigor'yev, Yuri, 'Antirossiyskiye Informatsionnyye Voyny' [Anti-Russian information wars]', *Informatsionnye voyny*, 4, 36 (2015), p. 6.
49. Lisichkin, Vladimir and Leonid Shelepin, *Tret'ya mirovaya informatsionno–psikhologicheskaya voyna* [The third world information–psychological war], (Moscow: Eskimo-Algoritm, 2003), p. 17.
50. Panarin, *Pervaya mirovaya informatsionnaya voyna*, p. 25.
51. Panarin and Panarina, *Informatsionnaya voyna i mir*, pp. 128–30.
52. For example: Panarin, *Pervaya mirovaya informatsionnaya voyna*; Panarin, *Informatsionnaya voyna i geopolitika*; Panarin, *Informatsionnyye voyna i kommunikatsii*.
53. Panarin, *Pervaya mirovaya informatsionnaya voyna*, p. 10.
54. For example: Panarin, *Pervaya mirovaya informatsionnaya voyna*; Panarin, *Informatsionnyye voyna i kommunikatsii*.
55. 'X' [George Kennan], 'The Sources of Soviet Conduct', *Foreign Affairs*, 25, 4 (1947), pp. 566–82.
56. Panarin, *Informatsionnyye voyna i kommunikatsii*, p. 118.
57. Ibid., p. 116.
58. Panarin and Panarina, *Informatsionnaya voyna i mir*, p. 4.
59. Panarin, *Informatsionnaya voyna i geopolitika*, pp. 244–5.
60. Panarin, *Informatsionnyye voyna i kommunikatsii*, pp. 200–1.
61. Ibid., p. 207.
62. Ibid., p. 186.
63. Ibid., p. 189.
64. For example: Kovalev, Victor and Sergey Malkov, 'Chto delat', chtoby ne raspast'sya kak SSSR?' [What should be done to not fall as the USSR?], *Informatsionnye voyny*, 3, 35 (2015), pp. 52–7; Tkachenko, Sergey, *Informatsionnaya voyna protiv Rossii* [Information war against Russia], Saint Petersburg: Piter, 2011; Lisichkin and Shelepin, *Tret'ya mirovaya informatsionno-psikhologicheskaya voyna*; Novikov, Vladimir, *Informatsionnoye oruzhiye: oruzhiye sovremennykh i budushchikh voyn* [Information-weapon: the weapon of contemporary and future wars], Moscow: Goryachaya Liniya-Telekom, 2011; Belyayev, Dmitry, *Razrukha v golovakh: Informatsionnaya voyna protiv Rossii* [The destruction in the heads: information war against Russia], Saint Petersburg: Piter, 2015.

5. THE RISE OF *GIBRIDNAYA VOYNA*

1. Belozerov, Vasily and Alexey Solov'ev, 'Gibridnaya Voyna v otechestvennom politicheskom i nauchnom diskurse' [Hybrid war in domestic political and acedemic discourse], *Vlast'*, 9 (2015), pp. 5–11.
2. For example: Klimenko, Darya and Gennady Nechaev, 'Armiya SSHA menyayet

strategiyu' [The US Army changes strategy], *Vzglyad: Delovaya Gazeta*, 24 June 2009; http://vz.ru/society/2009/6/24/300239.html (accessed: 28 November 2016).

3. Hoffman, Frank, 'Gibridnyye ugrozy: pereosmysleniye izmenyayushchegosya kharaktera sovremennykh konfliktov', *Geopolitika*, 21 (2013), pp. 45–63, translated from Hoffman, Frank, 'Hybrid Threats: Reconceptualising the Evolving Character of Modern Conflict', *Strategic Forum*, 240 (April 2009).
4. For example: Popov, Igor, 'Matritza vo'in sovremennoi epohi' [The matrix of wars in contemporary time], *Nezavisimoe Voennoe Obozrenie*, 22 March 2013; http://nvo.ng.ru/concepts/2013-03-22/7_matrix.html, (accessed: 28 November 2016); Sayapin, Vladislav, 'Sovremennyye vyzovy virtual'nykh voyn' [The contemporary challenges of virtual wars], *Istoricheskiye, filosofskiye, politicheskiye i yuridicheskiye nauki, kul'turologiya i iskusstvovedeniye. Voprosy teorii i praktiki*, 3, 12/38 (2013), pp. 180–5; Bel'kov, Oleg, '"Gibridnaya voyna": novaya real'nost' ili novoye slovo o starykh veshchakh?' ['Hybrid war': a new reality or a new word about old things?], *Bezopasnost' Yevrazii*, 1 (January–July 2015), pp. 231–4.
5. Belozerov and Solov'ev, 'Gibridnaya Voyna'.
6. Korybko, Andrew, *Hybrid Wars: The Indirect Adaptive Approach to Regime Change*, Moscow: People's Friendship University of Russia, 2015; Tsygankov, Pavel (ed.), 'Gibridnyye Voyny v khaotiziruyushchemsya mire XXI veka ['Hybrid wars' in the chaotic world of the twenty-first century], Moscow: Moscow University Press, 2015.
7. For example: Vladimirov, Alexander, 'Gosudarstvo, voyna i natsional'naya bezopasnost' Rossii' [State, war and the national security of Russia], *Prostranstvo i Vremya*, 1, 3 (2011), pp. 26–38; Pavlushenko, Zyzin and Ol'hovnik, 'Myatezhevoyna kak forma tsivilizatsionnogo stolknoveniya', pp. 13–19; Neklessa, Alexander, 'Gibridnaya Voyna: Oblik i palitra vooruzhennykh konfliktov v XXI veke' [Hybrid warfare: the armed conflict's shape and palette in the twenty-first century], *Ekonomicheskiye strategii*, 8 (2015), pp. 78–85; Zolotoryev, Pavel, 'Global'noe izmerenie voyny: novye podhody v XXI veke' [The global dimension of war: new approaches in the twenty-first century], *Rossiya v Global'noy Politike*, 8, 1 (January–February 2010), pp. 45–58.
8. See Chapter 3.
9. As stated on the journal's website: http://www.iwars.su/redkol (in Russian) (accessed: 28 November 2016).
10. See: Savinkin, Alexander (ed.), *Khochesh' mira, pobedi myatezhevoynu! Tvorcheskoye naslediye E. Messnera* [If you want peace, win subversion-war! The creative heritage of E. Messner], Moscow: Voyennyy Universitet, Russkiy Put', 2005; Messner, Evgeny, *Vsemirnaya Myatezhvoyna* [Worldwide subversion-war], Moscow: Zhukovskoye Pole, 2004; Savinkin, Alexander (ed.), *Russkoye zarubezh'ye: gosudarstvenno-patrioticheskaya i voyennaya mysl* [Russian emigrés: political-patriotic

and military thought], Moscow: Gumanitarnaya akademiya vooruzhennykh sil, 1994; Savinkin (ed.), *Dusha armii: Russkaya voyennaya emigratsiya o moral'no-psikhologicheskikh osnovakh rossiyskoy vooruzhennoy sily* [The soul of army: Russian military emigrés write on the moral-psychological foundations of Russian Armed Forces], Moscow: Voyennyy universitet, Nezavisimyy voyenno-nauchnyy tsentr 'Otechestvo i Voin', Russkiy put', 1997; Savinkin (ed.), *Voyennaya mysl' v izgnanii: tvorchestvo russkoy voyennoy emigratsii* [Military thought in exile: the oeuvre of Russian military emigrés], Moscow: Voyennyy Universitet, Russkiy Put', 1999.

11. For example: Domnin, Igor and Alexander Savinkin, 'Assimetrichnoe Voevanie' [Asymmetric warfare], *Otechestvennyye Zapiski*, 5 (2005); Biryukov, Sergey and Alexander Davydov, 'Konfliktnaya model' formirovaniya natsional'noy politicheskoy identichnosti: kontseptsiya "matezhevoyny" E. Messnera (na primere Bosnii)' [The conflict model of the formation of national identity: the concept of subversion-way by E. Messner (case study: Bosnia)]', *Politicheskiye Instituty i Protsessy*, 2 (2014), pp. 132–45; Nesterov, Aleksey, 'Netraditsionnyye ugrozy voyennoy bezopasnosti Rossiyskoy Federatsii v ramkakh modernizatsii NATO' [Non-traditional threats to the security of the Russian Federation due to the modernisation of NATO]', *Vlast'*, 5 (2013), pp. 186–8; Morozov, Evgeny, 'Predisloviye kommentatora' [Commentator's foreword], in Messner, *Vsemirnaya Myatezhvoyna*; Savinkin, Alexander, 'Groznaya Opasnost' Vsemirnoy Myatezhevoyny' [The mounting danger of the worldwide subversion-war], in Savinkin, *Khochesh' mira, pobedi myatezhevoynu*; Gracheva, Tatiana, 'Postizheniye voyny: myatezh kak sostoyaniye dushi, tsel' i sposob zavoyevaniya' [Comprehending war: subversion as a state of soul, a purpose and a way of conquest], in Messner, *Vsemirnaya Myatezhvoyna*; Pavlushenko, Mikhail, Vladimir Zyzin and Yury Ol'hovnik, 'Myatezhevoyna kak forma tsivilizatsionnogo stolknoveniya "Zapad-Vostok"' [Subversion-war as a form of the 'West–East' civilisationalal clash], *Obozrevatel'-Observer*, 5 (2007), pp. 13–19.
12. Hoffman, Frank, *Conflict in the 21st Century: The Rise of Hybrid Warfare*, Arlington: Potomac Institute for Policy Studies, 2007, p. 14.
13. Tsygankov, Pavel, 'Gibridnyye Voyny: ponyatiya, interpretatsii i real'nost' [Hybrid wars: definitions, interpretations and reality]', in Tsygankov, *'Gibridnyye Voyny'*, p. 21.
14. Svechin, Alexander, *Strategya* [Strategy], Moscow: Gosudarstvennoye voyennoye izdatel'stvo, 1926; Gareev, Makhmut, *Sovetskaya Voennaya Nauka* [The Soviet military science], Moscow: Izdatel'stvo Znanie, 1988; Serebryannikov, Vladimir, *Sotsiologiya voyny* [The sociology of war], Moscow: Nauchnyy mir, 1997.
15. Izmest'yev, P. and A. Messner, *Konspekt strategii* [The abstract of strategy], Saint-Petersburg: Tipo-Litografiya A.E. Landau, 1899.
16. Shnirelmann, Victor, 'U istokov voyny i mira' [At the source of war and peace], in Pershits, Abram, Yury Semenov and Victor Shnirelmann (eds), *Voyna i mir v*

ranney istorii chelovechestva [War and peace in early human history], vol. 1, Moscow: Institut etnologii i antropologii RAN, 1994, p. 56.

17. Petriy, Petr, 'K voprosu o metodologii analiza voyny kak sotsial'nogo fenomena' [The question of the methodology of the analysis of war as a social phenomenon], *Bezopasnost' Yevrazii*, 1 (2015), pp. 225–8; Kurochko, Michail, 'Panoplia voyny kak metodologiya analiza gibridnykh voyn' [The panoply of war as a methodology for the analysis of hybrid wars], *Bezopasnost' Yevrazii*, 1 (2015), pp. 228–31; Danilenko, Ignat, 'Problema klassifikatsii sovremennykh voyn' [The problem of the classification of contemporary wars], *Bezopasnost' Yevrazii*, 1 (2015), pp. 234–6; Bel'kov, Oleg, '"Gibridnaya Voyna": Novaya real'nost' ili novoye slovo o starykh veshchakh?', *Bezopasnost' Yevrazii*, 1 (2015), pp. 231–4.
18. See Chapter 2.
19. Nebrenchin, Sergey, 'Informatsionno-virtual'nyy kontekst sovremennykh gibridnykh voyn' [The contemporary challenges of virtual wars], *Bezopasnost' Yevrazii*, 1 (2015), p. 244.
20. Bartosh, Alexander, 'Kompleks podryvnykh tekhnologiy "Tsvetnaya Revolyutsiya—Gibridnaya Voyna" kak ugroza natsional'noy bezopasnosti Rossii' [The complex of subversive techniques: 'colour revolution–hybrid war', as a threat to Russia's national security], *Bezopasnost' Yevrazii*, 1 (2015), pp. 245–7; Manoylo, Andrey, 'Tekhnologii "tsvetnykh revolyutsiy" v sovremennykh proyavleniyakh "gibridnykh voyn"' [The techniques of 'colour revolutions' in contemporary expressions of 'hybrid wars'], in Tsygankov, *'Gibridnyye Voyny'*; Lobanov, Konstantin, 'Problemy obespecheniya natsional'noy bezopasnosti v protsesse protivodeystviya gibridnym i "tsvetnym" tekhnologiyam' [The problems of providing national security in the process of counteracting hybrid and 'colour' techniques], *Srednerusskiy vestnik obshchestvennykh nauk*, 11, 1 (2016), pp. 64–74; Korybko, *Hybrid Wars*.
21. Panarin, Igor, 'Gladiatory gibridnoy voyny' [The gladiators of hybrid war], *Ekonomicheskiye Strategii*, 2 (2016), p. 65.
22. Volodenkov Sergey, 'Informatsionnoye protivoborstvo kak sostavlyayushchaya sovremennykh gibridnykh voyn: rol' i osobennosti' [Informational struggle as a component of contemporary hybrid wars: its role and characteristics], in Tsygankov, *'Gibridnyye Voyny'*, p. 194.
23. Nevskaya, Tatyana, 'Informatsionnaya sostavlyayushchaya gibridnykh voyn' [The information component of hybrid wars], *Vestnik Moskovskogo universiteta. Seriya 18. Sotsiologiya i politologiya*, 4 (2015), pp. 281–4; Manoylo, Andrey, 'Gibridnaya voyna i tsvetnaya revolyutsiya: k voprosu o sootnoshenii ponyatiy' [Hybrid war and colour revolution: the question of the compatibility of definitions], in Bocharnikov, Igor (ed.), *Evolyutsiya form, metodov i instrumentov protivoborstva v sovremennykh konfliktakh* [The evolution of forms, methods and instruments of struggle in contemporary conflicts], Moscow: Ekon-Inform, 2015; Kharibin, Alexander, 'Myagkaya sila i gibridnaya voyna kak elementy strategii Rossii v geo-

politicheskikh konfliktakh' [Soft power and hybrid war as the elements of Russia's strategy in geopolitical conflicts], *Nauka Krasnoyar'ya*, 6, 23 (2015), pp. 70–79; Filimonov, Georgy, Nikita Danyk and Maxim Yurakov, *Perevorot* [Coup d'état], Saint-Petersburg: Piter, 2016.
24. Panarin, 'Gladiatory gibridnoy voyny', p. 65.
25. Panarin, Igor, *Gibridnaya voyna protiv Rossii, 1816–2016 gg* [Hybrid war against Russia, 1816–2016], Moscow: Goryachaya Liniya-Telekom, 2016, p. 3.
26. Tsygankov, 'Gibridnyye Voyny: ponyatiya, interpretatsii i real'nost', pp. 5–6.
27. Sivkov, Konstantin, oral presentation at the roundtable 'Armiya budushchego, vzglyad za gorizont' [The army of the future: view beyond the horizon], at the International Military–Technical Forum ARMY-2015, http://journal-otechestvo.ru/kruglyj-stol-armiya-buduschego/ (video; Russian) (accessed: 28 November 2016).
28. Raskin, Alexander, 'Setevyye Tekhnologii v gibridnoy voyne' [Network technologies in hybrid war], *Informatsionnyye voyny*, 1, 37 (2016), pp. 2–4; Neklessa, 'Gibridnaya Voyna'; Korybko, *Hybrid Wars*.
29. Stoletov, Oleg, '"Gibridnaya voyna" kak teoreticheskiy kontsept i instrument diskussionogo vliyaniya v sovremennoy mirovoy politike' ['Hybrid war' as a theoretical concept and an instrument of discursive influence in contemporary world politics], in Tsygankov, *'Gibridnyye Voyny'*; Achkasov, Valery, '"Gibridnyye viyny" kak instrument realizatsii strategii upravlyayemogo khaosa' ['Hybrid Wars' as an instrument to implement the strategy of controlled chaos], in Tsygankov, *'Gibridnyye Voyny'*; Krasnoslobodtsev, Vladimir, Alexander Raskin and Igor Tarasov, 'Gibridnaya voyna: ponyatiye, sushchnost', napravleniye protivodeystviya' [Hybrid war: definition, nature, the direction of counter-measures], *Strategicheskaya stabil'nost'*, 1, 74 (2016), pp. 6–9.
30. Budaev, Andrey, 'Gibridnyye voyny SSHA v gosudarstvakh Latinskoy Ameriki: podkhody i praktika primeneniya' [The hybrid wars of the United States in the states of Latin America: the methods and practice of its employment], in Tsygankov, *'Gibridnyye Voyny'*.
31. Manoylo, 'Tekhnologii "Tsvetnykh revolyutsiy"'.
32. Kuznetsov, Igor, 'Transformatsiya form voyny v sovremennoy politike: voyennoye i nevoyennoye protivostoyaniye na territorii Ukrainy 2014–2015 gg.' [The transformation of the forms of war in contemporary politics: military and non-military struggle on the territory of Ukraine, 2014–15], in Tsygankov, *'Gibridnyye Voyny'*; Kotlyar, Vladimir, 'K voprosu o "gibridnoy voyne" i o tom, kto zhe yeye vedet na Ukraine' [On the question of 'hybrid war' and who, after all, wages it in Ukraine], *Mezhdunarodnaya zhizn'*, 8 (2015), pp. 58–72.
33. See Chapter 2.
34. Gerasimov, Valery, presentation at the Russian Ministry of Defence's third Moscow Conference on International Security, Moscow, 23 May 2014, in Cordesman,

Anthony, 'Russia and the "Color Revolution": A Russian Military View of a World Destabilized by the US and the West', Centre for Strategic and International Studies, 24 May 2014.
35. Gerasimov, Valery, 'Tsennost' nauki v predvidenii: Novyye vyzovy trebuyut pereosmyslit' formy i sposoby vedeniya boyevykh deystviy' [The value of science is in foresight: new challenges demand rethinking the forms and methods of carrying out combat operations], *Voyenno-Promyshlennyy Kurier*, 8 (2013); http://vpk-news.ru/articles/14632 (accessed: 28 November 2016).
36. Smith, Paul, *On Political War*, Washington, DC: National Defense University Press, 1989, p. 3.
37. Gerasimov, Valery, 'Po opytu Sirii: Gibridnaya voyna trebuyet vysokotekhnologichnogo oruzhiya i nauchnogo obosnovaniya' [According to the experience in Syria: hybrid war requires high-tech weaponry and scientific foundations], *Voyenno-Promyshlennyy Kurier*, 9 (2016); http://vpk-news.ru/articles/29579, (accessed: 28 November 2016).
38. Bartles, Charles, 'Getting Gerasimov Right', *Military Review*, 96 (2016), p. 31.
39. Ibid.
40. For example: Filimonov, Georgy, 'Profilaktika ekstremizma' [Preventing extremism], oral presentation at the roundtable 'Rossiya v izmenyayushchemsya mire: vyzovy, opasnosti, ugrozy', 'Rossiya v izmenyayushchemsya mire: vyzovy, opasnosti, ugrozy' [Russia in a Changing World: Challenges, dangers and threats] at the International Military-Technical Forum ARMY-2016, (video), (Russian), http://journal-otechestvo.ru/armiya-2016-rossiya-v-izmenyayuschemsya-mire/, (accessed: 28 November 2016).
41. Ministry of Defence of the Russian Federation, *Voyennyy entsiklopedicheskiy slovar'* [Military encyclopaedic dictionary]; http://encyclopedia.mil.ru/encyclopedia/dictionary/list.htm (accessed: 28 November 2016).
42. Gerasimov, 'Tsennost' nauki v predvidenii'.

6. 'THE RUSSIANS ARE COMING'—THE POLITICISATION OF RUSSIAN HYBRID WARFARE

1. See Chapter 2.
2. Mansoor, Peter, 'Introduction', in Murray, Williamson and Peter Mansoor (eds), *Hybrid Warfare: Fighting Complex Opponents from the Ancient World to the Present*, Cambridge: Cambridge University Press, 2012, p. 2.
3. Ibid., p. 1.
4. Boot, Max, 'Countering Hybrid warfare', *Armed Conflict Survey*, 1 (2015), p. 11.
5. Murray, Williamson, 'Conclusions: What the Past Suggests', in Murray and Mansoor, *Hybrid Warfare*, p. 290.
6. Cox, Dan, Thomas Bruscino and Alex Ryan, 'Why Hybrid Warfare is Tactics Not

Strategy: A Rejoinder to "Future Threats and Strategic Thinking"', *Infinity Journal*, 2, 2 (Spring 2012), pp. 25, 26.
7. Strachan, Hew, *The Direction of War: Contemporary Strategy in Historical Perspective*, New York: Cambridge University Press, 2013, p. 82.
8. Scheipers, Sibylle, 'Winning War without Battles: Hybrid Warfare and Other Indirect Approaches in the History of Strategic Thought', in Renz, Bettina and Hanna Smith (eds), 'Russia and Hybrid Warfare: Going Beyond the Label', *Papers Aleksanteri*, 1 (2016), p. 51.
9. Cox, Bruscino, and Ryan, 'Why Hybrid Warfare is Tactics', p. 27.
10. See Chapter 2.
11. Renz, Bettina, 'Russia and "Hybrid Warfare"', *Contemporary Politics*, 22, 3 (2016), pp. 283–300.
12. Mihara, Robert, 'Beyond Future Threats: A Business Alternative to Threat-Based Strategic Planning', *Infinity Journal*, 2, 3 (Summer 2012), p. 25.
13. Friedman, Brett, 'Blurred Lines: The Myth of Guerrilla Tactics', *Infinity Journal*, 3, 4 (Winter 2014), p. 27.
14. Echevarria, Antulio, 'How Should We Think about "Gray-Zone" Wars?', *Infinity Journal*, 5, 1 (Fall 2015), pp. 16–20.
15. Freedman, Lawrence, 'Ukraine and the Art of Limited War', *Survival*, 56, 6 (2014), p. 11.
16. Kilcullen, David, *The Accidental Guerrilla: Fighting Small Wars in the Midst of a Big One*, London: Hurst & Company, 2009, Chapter 1.
17. North Atlantic Treaty Organisation, *BI-SC Input to a NEW Capstone Concept for the Military Contribution to Countering Hybrid Threats*, Brussels, 2010, pp. 2–3.
18. Aaronson, Michael, et al., 'NATO Countering the Hybrid Threat', *Prism*, 2, 4 (2011), pp. 111–24; Bachmann, Sascha-Dominik and Håkan Gunneriusson, 'Terrorism and Cyber Attacks as Hybrid Threats: Defining a Comprehensive Approach for Countering 21st-Century Threats to Global Peace and Security', *Journal of Terrorism and Security Analysis* (Spring 2014), pp. 26–36.
19. Aaronson, et al., 'NATO Countering the Hybrid Threat', p. 115.
20. Joint Irregular Warfare Center, 'Irregular Adversaries and Hybrid Threats, an Assessment: 2011', p. 24.
21. Bachmann, Sascha-Dominik and Håkan Gunneriusson, 'Hybrid Wars: The 21st-Century's New Threats to Global Peace and Security', *Scientia Militaria, South African Journal of Military Studies*, 43, 1 (2015), p. 79.
22. Bachmann and Gunneriusson, 'Terrorism and Cyber Attacks as Hybrid Threats', p. 28.
23. Bachmann and Gunneriusson, 'Hybrid Wars: The 21st-Century's New Threats to Global Peace', p. 78; see also: Bachmann and Gunneriusson, 'Terrorism and Cyber Attacks as Hybrid Threats'; Bachmann, Sascha-Dominik and Gerhard Kemp, 'Aggression as "Organized Hypocrisy?": How the War on Terrorism and Hybrid

Threats Challenge the Nuremberg Legacy', *Windsor Year Book of Access to Justice*, 30, 1 (2012), pp. 233–252.

24. For example: Wither, James, 'Making Sense of Hybrid Warfare', *Connections: The Quarterly Journal*, 15, 2 (2016), pp. 73–87; Monaghan, Andrew, 'The "War" in Russia's "Hybrid Warfare"', *Parameters*, 45, 4 (Winter 2015–16), pp. 65–74; Renz, 'Russia and "Hybrid Warfare"'.

25. 'Editor's Introduction: Complex Crises Call for Adaptable and Durable Capabilities', *Military Balance*, 115, 1 (2015), p. 5.

26. 'Research Publications', NATO Defence College, http://www.ndc.nato.int/research/research.php?icode=0 (accessed: 15 January 2017).

27. Breedlove, Philip, 'Foreword', in Lasconjarias, Guillaume and Jeffrey Larsen (eds), *NATO's Response to Hybrid Threats*, Rome: NATO Defence College, 2015, p. xxii.

28. Ruiz Palmer, Diego, 'Back to the Future? Russia's Hybrid Warfare, Revolutions in Military Affairs, and Cold War Comparisons', Research Paper no. 120, NATO Defence College, Rome, October 2015, p. 2.

29. Reisinger, Heidi, 'Does Russia Matter? Purely Political Relations Are Not Enough in Operational Times', NDC Conference Report, NATO Defence College, Rome, 31 January 2014, p. 3.

30. Reisinger, Heidi and Aleksandr Golts, 'Russian Hybrid Warfare: Waging War below the Radar of Traditional Collective Defence', Research Paper no. 105, NATO Defence College, Rome, November 2014, p. 3.

31. 'Hybrid War: Hybrid Response?', *NATO Review Magazine* (Video), http://www.nato.int/docu/review/2014/Russia-Ukraine-Nato-crisis/Russia-Ukraine-crisis-war/EN/index.htm, 1 July 2014 (accessed: 19 January 2017).

32. Ibid.

33. Van Kappen, Frank, quoted in Kornienko, Sofia, 'Pidjak rvetsya po shvam' [Jacket is torn at the seams], *Radio Svoboda*, 26 April 2014; http://www.svoboda.org/a/25362031.html (accessed: 7 February 2017).

34. Landler, Mark and Michael Gordon, 'NATO Chief Warns of Duplicity by Putin on Ukraine', *New York Times*, 8 July 2014; https://www.nytimes.com/2014/07/09/world/europe/nato-chief-warns-of-duplicity-by-putin-on-ukraine.html?_r=0 (accessed: 19 January 2017).

35. Padurariu, Dinu, Nicolae Cretu and Vasile Onesimuc, 'The Hybrid Warfare: Threats to the Security of the states at the NATO East Border', *International Scientific Conference 'Strategies XXI'*, suppl. Command and Staff Faculty, 2 (2014), p. 6.

36. 'Hybrid War: Hybrid Response?'

37. Paul King, e-mail message to author, 25 January 2017.

38. Hoffman, Frank, 'On Not-So-New Warfare: Political Warfare vs Hybrid Threats', *War on the Rocks*, 28 July 2014; https://warontherocks.com/2014/07/on-not-so-new-warfare-political-warfare-vs-hybrid-threats/ (accessed: 19 January 2017).

39. Ibid.
40. North Atlantic Treaty Organisation, Wales Summit Declaration: Issued by the Heads of State and Government Participating in the Meeting of the North Atlantic Council in Wales from 4 to 5 September 2014, 5 September, 2014, p. 1.
41. Ibid., p. 3.
42. Vandiver, John, 'SACEUR: Allies Must Prepare for Russia "Hybrid war"', *Stars and Stripes*, 4 September, 2014; http://www.stripes.com/news/saceur-allies-must-prepare-for-russia-hybrid-war-1.301464 (accessed: 19 January 2017).
43. For example: Gonchar, Mykhailo, Andriy Chubyk and Oxana Ishshuk, 'Hybrid War in Eastern Europe: Non-military Dimension', *International Issues & Slovak Foreign Policy Affairs*, 23, 3–4 (2014), pp. 27–36; Popescu, Alba, 'Observations Regarding the Actuality of the Hybrid War: Case Study; Ukraine', *Strategic Impact*, 4 (2014), pp. 118–33; Ionita, Craisor-Constantin, 'Is Hybrid Warfare Something New?', *Strategic Impact*, 4 (2014), pp. 61–71; Schmidt, Nicola, 'Neither Conventional War, nor a Cyber War, but a Long-Lasting and Silent Hybrid War', *Obrana a Strategie*, 2 (2014), pp. 73–85; Necula, Dragos Alexandru, 'The Hybrid War: New Name; Old Tactics', paper presented at the International Scientific Conference 'Strategies XXI', Bucharest, 2014; Padurariu, Dinu, Cretu Nicolae and Vasile Onesimiuc, 'The Hybrid Warfare Threats to the Security of the States at the NATO East Border', paper presented at the International Scientific Conference 'Strategies XXI', Bucharest, 2014.
44. For example: Nistor, Florin, '"Insidious Operations": A New Concept of Waging War for Russia?', paper presented at the International Scientific Conference 'Strategies XXI', Bucharest, 2014; Ionita, 'Is Hybrid Warfare Something New?'
45. Nistor, '"Insidious Operations"'.
46. For example: Ionita, 'Is Hybrid Warfare Something New?'; Popescu, 'Observations Regarding the Actuality of the Hybrid War'; Necula, 'Hybrid War: New Name; Old Tactics'; Schmidt, 'Neither Conventional War, nor a Cyber War'.
47. For example: Nistor, '"Insidious Operations"'.
48. Paul King, e-mail message to author, 25 January 2017.
49. Popescu, 'Observations Regarding the Actuality of the Hybrid War'.
50. Necula, 'The Hybrid War: New Name; Old Tactics'.
51. Ibid.
52. Padurariu, Dinu, Cretu, Nicolae and Vasile Onesimiuc, 'The Hybrid Warfare Threats to the Security of the States at the NATO East Border', paper presented at the International Scientific Conference 'Strategies XXI', Bucharest, 2014.
53. For example: Iancu, Niculae, et al. (eds), *Countering Hybrid Threats: Lessons Learned from Ukraine*, Amsterdam: IOS Press, 2016; Banasik, Miroslaw, 'How to Understand the Hybrid War', *Securitologia*, 1 (2015), pp. 19–34; Necula, Dragos-Alexandru, 'NATO and UU Views on Vulnerabilities to Hybrid Warfare Actions', paper presented at the International Scientific Conference 'Strategies XXI',

Bucharest, 2015; Popescu, Nicu, 'Hybrid Tactics: Neither New nor Only Russian', *EUISS Issue Alert*, 4 (2015); Pindják, Peter, 'Optimizing Armed Forces Capabilities for Hybrid Warfare: New Challenge for Slovak Armed Forces', *INCAS Bulletin*, 7, 3 (2015), pp. 191–8; Munteanu, Ravzan, 'Hybrid Warfare: The New Form of Conflict at the Beginning of the Century', *Strategic Impact*, 56 (2015), pp. 19–26.

54. For example: Rácz, András, 'Russia's Hybrid War in Ukraine: Breaking the Enemy's Ability to Resist', *FIFA Report*, 43 (June 2015); Finnish Defence Research Agency (FDRA), 'On the Concept of Hybrid Warfare', Research Bulletin no. 1 (2015); Cederberg, Aapo and Pasi Eronen, 'How Can Societies Be Defended against Hybrid Threats?', *Strategic Security Analysis*, 9 (September 2015).

55. For example: Hunter, Eve and Piret Penik, 'The Challenges of Hybrid Warfare', International Centre for Defence and Security, Tallinn, April 2015; Maigre, Merle, 'Nothing New in Hybrid Warfare: The Estonian Experience and Recommendations for NATO', Policy Brief, German Marshall Fund of the United States, February 2015; Berzins, Janis, 'Russia's New Generation Warfare in Ukraine: Implications for Latvian Defense', Policy Paper no. 2, National Defence Academy of Latvia, Centre for Security and Strategic Research, April 2014.

56. For example: 'Editor's Introduction: Complex Crises Call for Adaptable and Durable Capabilities', *Military Balance*, 115, 1 (2015); Charap, Samuel, 'The Ghost of Hybrid War', *Survival*, 57, 6 (December 2015–January 2016), pp. 51–8; Freedman, 'Ukraine and the Art of Limited War'; Barry, Ben, 'Hybrid Warfare: The Conflict in Ukraine and the War against ISIS', discussion meeting at the International Institute for Strategic Studies (IISS), 19 October 2015 (video); https://www.iiss.org/en/events/events/archive/2015-f463/october-a702/hybrid-warfare-the-conflict-in-ukraine-and-the-war-against-isis-1079 (accessed: 9 February 2017).

57. For example: Kofman, Michael and Matthew Rojansky, 'A Closer Look at Russia's "Hybrid War"', *Kennan Cable*, no. 7 (April 2015); Neville, Seth, 'Russia and Hybrid Warfare: Identifying Critical Elements in Successful Applications of Hybrid Tactics', Monterey: Naval Postgraduate School, 2015; Bartkowski, Maciej, 'Nonviolent Civilian Defense to Counter Russian Hybrid Warfare', Johns Hopkins University Center for Advanced Governmental Studies, March 2015.

58. 'Research Publications', NATO Defence College.

59. Lasconjarias and Larsen, *NATO's Response to Hybrid Threats*, p. xxvii.

60. Iancu et al., *Countering Hybrid Threats*.

61. Rácz, 'Russia's Hybrid War in Ukraine'.

62. Ibid., p. 42. For the article on 'full-spectrum conflict', see: Jonsson, Oscar and Robert Seely, 'Russian Full-Spectrum Conflict: An Appraisal After Ukraine', *Journal of Slavic Military Studies*, 28, 1 (2015), pp. 1–22.

63. Freedman, 'Ukraine and the Art of Limited War', pp. 10–11.

64. Gerasimov, Valery, 'Tsennost' nauki v predvidenii: Novyye vyzovy trebuyut pereos-

myslit' formy i sposoby vedeniya boyevykh deystviy' [The value of science is in foresight: new challenges demand rethinking the forms and methods of carrying out combat operations], *Voyenno-Promyshlennyy Kurier*, 8 (2013); http://vpknews.ru/articles/14632 (accessed: 28 November 2016).

65. Bartles, 'Getting Gerasimov Right', *Military Review*. See also Chapter 7.
66. Gerasimov, Valery, presentation at the Russian Ministry of Defense's third Moscow Conference on International Security, Moscow, 23 May 2014, in Cordesman, 'Russia and the "Colour Revolution"'; Gerasimov, 'Tsennost' nauki v predvidenii'.
67. Chekinov, Sergey and Sergey Bogdanov, 'Assimetrichnyye deystviya po obespecheniyu voyennoy bezopasnosti rossii' [Asymmetric actions that assure Russia's military security], *Voennaya Mysl'*, 3 (2010), pp. 13–22; Chekinov, Sergey and Sergey Bogdanov, 'Vliyaniye nepryamykh deystviy na kharakter sovremennoy voyny nachal'nogo perioda XXI stoletiya' [The influence of indirect actions on the character of contemporary war during the initial period of the twenty-first century], *Voennaya Mysl'*, 6 (2011), pp. 3–13; Chekinov, Sergey and Sergey Bogdanov, 'Nachal'nyye periody voyn i ikh vliyaniye na podgotovku strany k voyne budushchego' [The initial periods of wars and their influence on preparing the country for future wars], *Voennaya Mysl'*, 11 (2012), pp. 14–27; Chekinov, Sergey and Sergey Bogdanov, 'O kharaktere i soderzhanii voyny novogo pokoleniya' [The nature and the content of a new-generation war], *Voennaya Mysl'*, 10 (2013), pp. 13–24.
68. For example: Johnson, Dave, 'Russia's Approach to Conflict: Implications for NATO's Deterrence and Defence', Research Paper no. 111, NATO Defence College, Rome, April 2015; Berzins, 'Russia's New Generation Warfare in Ukraine'; Maigre, 'Nothing New in Hybrid Warfare'; Banasik, 'How to Understand the Hybrid War'; Ruiz Palmer, 'Back to the Future? Russia's Hybrid Warfare'.
69. Adamsky, Dima, 'Cross-Domain Coercion: The Current Russian Art of Strategy', *Prolifiration Papers*, 54 (November 2015), p. 21.
70. Praks, Henrik, 'Hybrid or Not: Deterring and Defeating Russia's Way of Warfare in the Baltics; The Case of Estonia', Research Paper no. 124, NATO Defence College, Rome, October 2015, pp. 10–11; see also: Barbu, Florin-Marian, 'Considerations Concerning Hybrid War', *Strategic Impact*, 2 (2015), pp. 50–6; Radulescu, Marian, 'Counter-Hybrid Warfare: Developments and Ways of Counteracting Hybrid Threats/War', paper presented at the International Scientific Conference 'Strategies XXI', Bucharest, 2015; Johnson, 'Russia's Approach to Conflict'; Maigre, 'Nothing New in Hybrid Warfare'; Popescu, 'Hybrid Tactics: Neither New nor Only Russian'; Ruiz Palmer, 'Back to the Future? Russia's Hybrid Warfare'.
71. NATO, 'NATO's Readiness Action Plan: Fact Sheet', December 2014; http://www.nato.int/nato_static_fl2014/assets/pdf/pdf_2014_12/20141202_141202-facstsheet-rap-en.pdf (accessed: 15 February 2017).
72. Supreme Headquarters Allied Powers Europe (SHAPE), 'Fact Sheet: NATO

Response Force (NRF)', January 2016; http://www.shape.nato.int/page349011837 (accessed: 15 February 2017).

73. Johnson, 'Russia's Approach to Conflict', p. 12; see also: Henrik, 'Hybrid or Not: Deterring and Defeating Russia's Way of Warfare'; Barbu, 'Considerations Concerning Hybrid War'; Radulescu, 'Counter-Hybrid Warfare'; Maigre, 'Nothing New in Hybrid Warfare'; Popescu, 'Hybrid Tactics: Neither New nor Only Russian'; Ruiz Palmer, 'Back to the Future? Russia's Hybrid Warfare'.

74. Kofman, Michael, 'Russian Hybrid Warfare and Other Dark Arts', *War on the Rocks*, 11 March 2016; https://warontherocks.com/2016/03/russian-hybrid-warfare-and-other-dark-arts/ (accessed: 15 February 2017).

75. This is not to say, however, that there was no further work promoting the applicability of the concept to Russia's actions. See, for example: Giegerich, Bastian, 'Hybrid Warfare and the Changing Character of Conflict', *Connections: The Quarterly Journal*, 2 (2016), pp. 65–72; Banasik, Miroslaw, 'Russia's Hybrid War', *Science & Military Journal*, 11, 2 (2016), pp. 3–47; Kiesewetter, Roderich and Ingmar Zielke, 'Permanent NATO Deployment Is Not the Answer to European Security', *European View*, 15 (2016), pp. 37–45; Dykyi, Evgen, *The 'Hybrid' War of Russia: Experience of Ukraine for the Baltic States*, Vilnius: General Jonas Žemaitis Military Academy of Lithuania, 2016.

76. For example: Thomas, Timothy, 'The Evolution of Russian Military Thought: Integrating Hybrid, New-Generation, and New-Type Thinking', *Journal of Slavic Military Studies*, 29, 4 (2016), pp. 554–75; Schadlow, Nadia, 'The Problem with Hybrid Warfare', *War on the Rocks*, 9 April 2015; http://warontherocks.com/2015/04/the-problem-with-hybrid-warfare/ (accessed: 15 February 2017); Raitasalo, Jyri, 'Hybrid Warfare: Where's the Beef?', *War on the Rocks*, 23 April 2015; http://warontherocks.com/2015/04/hybrid-warfare-wheres-the-beef/ (accessed: 15 February 2017).

77. Renz, Bettina and Hanna Smith (eds), 'After "Hybrid Warfare", What Next? Understanding and Responding to Contemporary Russia', Publications of the Government's analysis, assessment and research activities 44/2016, Helsinki, 2016, pp. 8–9.

78. Galeotti, Mark, *Hybrid War or Gibridnaya Voina? Getting Russia's Non-Linear Military Challenge Right*, Prague: Mayak Intelligence, 2016, p. 76.

79. For example: Giles, Keir, 'Russia's "New" Tools for Confronting the West: Continuity and Innovation in Moscow's Exercise of Power', Research Paper, Chatham House, London, March 2016; Monaghan, Andrew, 'The "War" in Russia's "Hybrid Warfare"', *Parameters*, 45, 4 (Winter 2015–16), pp. 65–74; Renz, 'Russia and "Hybrid Warfare"'; Galeotti, *Hybrid War or Gibridnaya Voina?*; Renz and Smith, 'Russia and Hybrid Warfare'; Renz and Smith, 'After "Hybrid Warfare", What Next?'.

80. For example: Persson, Gudrun (ed.), *Russian Military Capability in a Ten-Year*

Perspective: 2016, Stockholm: Swedish Defence Research Agency, 2016; Monaghan, Andrew, 'Russian State Mobilization: Moving the Country on to a War Footing', Research Paper, Chatham House, London, May 2016; Renz, Bettina, 'Why Is Russia Rebuilding Its Conventional Military Power?', in Renz and Smith, 'After "Hybrid Warfare", What Next?'.

81. For example: Thornton, Rod, 'Russian "Hybrid Warfare" and the National Defence Management Centre (NTsUO)', in Renz and Smith, 'After "Hybrid Warfare", What Next?'; Galeotti, *Hybrid War or Gibridnaya Voina?*

82. For example: Thomas, Timothy, 'Dialectical Versus Empirical Thinking: Ten Key Elements of the Russian Understanding of Information Operations', *Journal of Slavic Military Studies*, 11, 1 (1998), pp. 40–62; Light, Margot, 'Information War', *World Today*, 56, 2 (2000), pp. 10–12; Herd, Graeme, 'The "Counter-Terrorist Operation" in Chechnya: "Information Warfare" Aspects', *Journal of Slavic Military Studies*, 13, 4 (2000), pp. 57–83. For a comprehensive review of the literature published during the 2000s see: Heickerö, Roland, *Emerging Cyber Threats and Russian Views on Information Warfare and Information Operations*, Stockholm: Swedish Defence Research Agency, 2010.

83. For example: Darczewska, Jolanta, 'The Anatomy of Russian Information Warfare: The Crimean Operation, A Case Study', *Point of View*, 42 (2014); Snegovaya, Maria, 'Putin's Information Warfare in Ukraine: Soviet Origins of Russia's Hybrid Warfare', *Russia Report*, 1 (September 2015); Thornton, Rod, 'The Changing Nature of Modern Warfare: Responding to Russian Information Warfare', *RUSI Journal*, 160, 4 (2015), pp. 40–8; Lucas, Edward and Peter Pomeranzev, *Winning the Information War: Technique and Counter-strategies to Russian Propaganda in Central and Eastern Europe*, Washington, DC: Center for European Policy Analysis, 2016; Kalinina, Ekaterina, 'Narratives of Russia's "Information Wars"', *Politics in Central Europe*, 12, 1 (2016), pp. 147–65.

84. Ibid.

85. The NATO Strategic Communications Centre of Excellence; http://www.stratcomcoe.org/publications (accessed: 16 February 2017).

86. For example: Giles, Keir, 'Russia and Its Neighbours: Old Attitudes, New Capabilities', in Geers, Kenneth, (ed.), *Cyber War in Perspective: Russian Aggression against Ukraine*, Tallinn: NATO CCD COE Publications, 2015; Giles, Keir, 'Russian Toolkit', in Giles, Keir, et. al., 'The Russian Challenge', Chatham House Report, Chatham House, London, June 2015; Giles, 'The Next Phase of Russian Information Warfare', NATO Strategic Communications Centre of Excellence, Riga, 2016; Giles, *Handbook of Russian Information Warfare*, Rome: NATO Defence College, 2016; Giles, 'The Next Phase of Russian Information Warfare', in Renz and Smith, 'Russia and Hybrid Warfare: Going Beyond the Label'.

87. Galeotti, *Hybrid War or Gibridnaya Voina?*, p. 4.

88. Giles, 'Russian Toolkit', pp. 48–9.

89. Giles, 'Next Phase of Russian Information Warfare', p. 4.
90. Galeotti, Mark, 'Hybrid, Ambiguous, and Non-linear? How New Is Russia's "New Way of War"?', *Small Wars & Insurgencies*, 27, 2 (2016), p. 282.
91. Galeotti, *Hybrid War or Gibridnaya Voina?*, p. 78.
92. Giles, *Handbook of Russian Information Warfare*, p. 60.
93. Giles, 'Russia's "New" Tools for Confronting the West', p. 59.
94. Inkster, Nigel, 'Information Warfare and the US Presidential Elections', *Survival*, 58, 5 (2016), pp. 23–32;
95. For example: Deyermond, Ruth, 'Russia's Trump Card? The Prospects for Russia–US Relations after the Election of Donald Trump', *Russian Analytical Digest*, 194 (2016), pp. 2–4; Zhemukhov, Sufian, 'The Kremlin's New Policy towards the US', *Russian Analytical Digest*, 194 (2016), pp. 5–7; Orttung, Robert, 'The Role of Russia in the US Presidential Elections', *Russian Analytical Digest*, 194 (2016), pp. 8–10; Inkster, 'Information Warfare and the US Presidential Elections'.
96. Intelligence Community Assessment, 'Assessing Russian Activities and Intentions in Recent US Elections', Washington, 6 January 2017, p. ii.
97. Behnke, Andreas, *NATO's Security Discourse after the Cold War*, London: Routledge, 2013, p. 3.
98. Rynning, Sten, *NATO in Afghanistan: The Liberal Disconnect*, Stanford: Stanford University Press, 2012; Behnke, *NATO's Security Discourse*, Chapter 10.
99. Larrabee, Stephen, *NATO and the Challenges of Austerity*, Santa Monica: Rand Corporation, 2012, p. 95.
100. NATO Strategic Communications Centre of Excellence, 'History'; http://www.stratcomcoe.org/history (accessed: 22 February 2017).
101. North Atlantic Treaty Organisation, 'Wales Summit Declaration', p. 3.
102. Kofman, 'Russian Hybrid Warfare and Other Dark Arts'.
103. Männik, Erik, 'Small States: Invited to NATO; Able to Contribute?', *Defense & Security Analysis*, 20, 1 (2004), p. 29; See also: Szulda, Robert and Marek Madej, (eds), *Newcomers No More? Contemporary NATO and the Future of Enlargement from the Perspective of 'post-Cold War' Members*, Warsaw: International Relations Research Institute, 2015.
104. Grybauskaitė, Dalia, quoted in Adomaitis, Nerijus, 'Lithuania Awaits NATO "Insurance Plan" on Russia', Reuters, 11 March 2010; http://in.reuters.com/article/idINIndia-46841820100311 (accessed: 22 February 2017).
105. Shlapak, David and Michael Johnson, 'Reinforcing Deterrence on NATO's Eastern Flank: War-Gaming the Defence of the Baltics', RAND Corporation, Santa Monica, 2016.
106. North Atlantic Treaty Organisation, 'Wales Summit Declaration', pp. 1–2.
107. Freedman, Lawrence, 'Academics and Policy-Making: Rules of Engagement', *Journal of Strategic Studies* (2017), p. 2.
108. The initial debate during 2014 and 2015.
109. For example, those expressed by Galeotti and Giles.

110. For example: Freedman, Lawrence, 'Stop Overestimating the Threat Posed by Russia's "New" Form of Warfare', World Economic Forum, 4 January 2017; https://www.weforum.org/agenda/2017/01/stop-overestimating-the-threat-posed-by-russia-s-new-form-of-warfare/ (accessed: 22 February 2017); Kofman, 'Russian Hybrid Warfare and Other Dark Arts'.
111. Freedman, 'Academics and Policy-Making', p. 3.
112. For example: Kanter, Jake and Adam Bienkov, 'Labour MPs Think the Government Is Hiding Info about Russia Interfering with Brexit', Business Insider UK, 23 February 2017; http://uk.businessinsider.com/labour-mp-ben-bradshaw-suspicious-russian-interference-brexit-2017-2 (accessed: 25 February 2017).
113. For example: Wagstyl, Stefan, 'German Politics: Russia's Next Target? Officials Fear Moscow May Interfere in This Year's Election to Benefit the Hard-Right', *Financial Times*, 27 January 2017; https://www.ft.com/content/31a5758c-e3d8-11e6-9645-c9357a75844a (accessed: 25 February 2017).
114. Intelligence Community Assessment, 'Assessing Russian Activities'.
115. European Union Institute for Security Studies (EUISS), 'EU Strategic Communications with a View to Counteracting Propaganda', Paris, 19 May 2016, p. 29.
116. European Parliament, 'European Parliament Resolution of 23 November 2016 on EU Strategic Communication to Counteract Propaganda against it by Third Parties (2016/2030(INI))', 23 November 2016, Brussels, p. 5.
117. Intelligence Community Assessment, 'Assessing Russian Activities', p. 1.
118. Priebus, Reince quoted in Weber, Joseph, 'Priebus Says US Intel Officials Call Campaign-Russia Story "Garbage," Tries to End Controversy', Fox News, 19 February 2017; http://www.foxnews.com/politics/2017/02/19/priebus-says-us-intel-officials-call-campaign-russia-story-garbage-tries-to-end-controversy.html (accessed: 25 February 2107).
119. Adamsky, 'Cross-Domain Coercion', p. 21.
120. Thomas, 'Evolution of Russian Military Thought', p. 574.
121. Obama, Barack, 'Speech to the United Nations General Assembly', 28 September 2015; https://www.nytimes.com/2015/09/29/world/americas/president-obamas-speech-to-the-united-nations-general-assembly-2015.html?_r=0 (accessed; 25 February 2017).
122. Kofman, 'Russian Hybrid Warfare and Other Dark Arts'.
123. European Parliament, 'European Parliament Resolution', p. 5.
124. Intelligence Community Assessment, 'Assessing Russian Activities', p. ii.

7. '*GIBRIDNAYA VOYNA* AGAINST US IS COMING': THE POLITICISATION OF WESTERN *GIBRIDNAYA VOYNA*

1. See 'Information about authors', *Voennaya Mysl'*, 1 (2017), pp. 94–5. For information

about the Centre for Military and Strategic Studies of the General Staff of the Russian Federation's Armed Forces see: Chekinov, Sergey, 'Tsentr voenno-strategicheskikh issledovanii' General'nogo shtaba Vooruzhennykh Sil Rossii'skoi' Federatsii: istoriia i sovremennost' (k 25-letiiu sozdaniia)' [The Centre for Military and Strategic Studies of the General Staff of the Russian Federation Armed Forces (the twenty-fifth anniversary)], *Voennaya Mysl'*, 1 (2010), pp. 3–5.

2. Gerasimov, Valery, 'Tsennost' nauki v predvidenii'; Gerasimov, presentation at the Russian Ministry of Defence's third Moscow Conference on International Security, Moscow, 23 May 2014, in Cordesman, Anthony, 'Russia and the "Colour Revolution"'.

3. Orekhov, Oleg and Sergey Chekinov, 'Kharakternyye osobennosti vooruzhennoy bor'by v voynakh i konfliktakh poslednego desyatiletiya' [The characterising features of the armed struggle in the wars and conflicts of the last decade], *Voennaya Mysl'*, 10 (2004), p. 13.

4. Chekinov, Sergey and Sergey Bogdanov, 'Evoliutsiia sushchnosti i soderzhaniia poniatiia "voi'na" v XXI stoletii' [The evolution of the nature and the content of the Concept of 'war' in the twenty-first century], *Voennaya Mysl'*, 1 (2017), p. 32; see also Chekinov, Sergey and Sergey Bogdanov, 'Voennoe iskusstvo na nachal'nom etape XXI stoletiya: problemy i suzhdeniya' [Military art at the beginning of the twenty-first century: problems and opinions], *Voennaya Mysl'*, 1 (2015), pp. 32–43.

5. Chekinov and Bogdanov, 'Evoliutsiia sushchnosti i soderzhaniia', p. 36; see also: Gareev, Makhmut, 'Voyna i voennaya nauka na sovremennom etape' [Contemporary war and military science], *Voenno-Promyshlennyy Kur'er*, 13 (2013); http://vpk-news.ru/sites/default/files/pdf/VPK_13_481.pdf (accessed: 1 March 2017).

6. Orekhov and Chekinov, 'Kharakternyye osobennosti vooruzhennoy bor'by', p. 16.

7. Chekinov, Sergey, 'Prognozirovanie tendentsii' voennogo iskusstva v nachal'nom periode XXI veka' [Forecasting the trends in the art of war in the initial period of the twenty-first century], *Voennaya Mysl'*, 7 (2010), p. 25.

8. Chekinov, Sergey and Sergey Bogdanov, 'Nachal'nye periody voyn I ikh vliyanie na podgotovku strany k voyne budushchgo' [The initial periods of wars and their influence on preparing the country for future wars], *Voennaya Mysl'*, 11 (2012), pp. 23–4.

9. Kokoshin, Andrey, *Politologia i sotziologia voyennoy strategii* [The politics and sociology of military strategy], Moscow: Lenand, 2016, pp. 502–3.

10. Volkovsky, Nikolay, *Istroiya informatzionnykh voyn* [The history of information wars], vol. 2, Saint Petersburg: Poligon, 2013, p. 660.

11. The vast majority of the cited articles by Chekinov and Bogdanov start with an analysis of the geopolitical transformations.

12. Chekinov and Bogdanov, 'Evoliutsiia sushchnosti i soderzhaniia', p. 35.

13. See Chapter 6.

14. Chekinov, Sergey and Sergey Bogdanov, 'O kharaktere i soderzhanii voyny novogo

pokoleniya' [The nature and content of a new-generation war], *Voennaya Mysl'*, 10 (2013), pp. 13–24.

15. For example: Johnson, Dave, 'Russia's Approach to Conflict: Implications for NATO's Deterrence and Defence', Research Paper no. 111, NATO Defence College, Rome, April 2015; Berzins, Janis, 'Russia's New Generation Warfare in Ukraine: Implications for Latvian Defense', Policy Paper no. 2, National Defence Academy of Latvia, Centre for Security and Strategic Research, April 2014; Maigre, Merle, 'Nothing New in Hybrid Warfare: The Estonian Experience and Recommendations for NATO', Policy Brief, German Marshall Fund of the United States, February 2015; Ruiz Palmer, Diego, 'Back to the Future?' Russia's Hybrid Warfare, Revolutions in Military Affairs, and Cold War Comparisons', Research Paper no. 120, NATO Defence College, Rome, October 2015; Galeotti, Mark, *Hybrid War or Gibridnaya Voina? Getting Russia's Non-Linear Military Challenge Right*, Prague: Mayak Intelligence, 2016.
16. Chekinov and Bogdanov, 'O kharaktere i soderzhanii', p. 17.
17. Ibid., pp. 19–20.
18. Chekinov and Bogdanov, 'Evoliutsiia sushchnosti i soderzhaniia', p. 39.
19. Gareev, 'Voyna i voennaya nauka na sovremennom etape'; see also: Gareev, Makhmut, 'Struktura I osnovnoe soderzhanie novoy voennoy doktriny' [The structure and content of the new military doctrine], *Voenno-Promyshlennyy Kur'er*, 3 (2007), http://www.vpk-news.ru/sites/default/files/pdf/issue_169.pdf (accessed: 1 March 2017).
20. Chekinov, Sergey and Sergey Bogdanov, 'Strategicheskoe sderzhivanie i nazhional'naya bezopasnost' Rossii na soveremennom etape' [The strategic deterrence and national security of Russia in the modern age], *Voennaya Mysl'*, 3 (2012), p. 15.
21. Ibid., p. 16.
22. Ibid., p. 12.
23. Chekinov, Sergey and Sergey Bogdanov, 'Vliianie nepriamykh dei'stvii' na kharakter sovremennoi' voi'ny' [The influence of indirect actions on the character of contemporary war], *Voennaya Mysl'*, 6 (2011), p. 7.
24. See Chapters 3–4.
25. Chekinov and Bogdanov, 'Vliianie nepriamykh dei'stvii'', p. 8.
26. Chekinov and Bogdanov, 'O kharaktere i soderzhanii'.
27. Chekinov and Bogdanov, 'Vliianie nepriamykh dei'stvii'', pp. 7, 10.
28. Chekinov and Bogdanov, 'Evoliutsiia sushchnosti i soderzhaniia', p. 40.
29. Ibid., p. 41.
30. Ibid.
31. Chekinov and Bogdanov, 'Nachal'nye periody', p. 26.
32. Chekinov and Bogdanov, 'Strategicheskoe sderzhivanie', p. 18.
33. Chekinov and Bogdanov, 'Vliianie nepriamykh dei'stvii'', p. 13.

34. Chekinov, Sergey and Sergey Bogdanov, 'Voennaya strategiya: vzglyad v budushcheye' [Military strategy: a peak into the future], *Voennaya Mysl'*, 11 (2016), p. 3.
35. Chekinov, Sergey and Sergey Bogdanov, 'Asimmetrichnye dei'stviia po obespecheniiu voennoi' bezopasnosti Rossii' [Asymmetric actions that ensure the military security of Russia], *Voennaya Mysl'*, 3 (2010), p. 17.
36. Ibid., p. 19.
37. Ibid., pp. 20–1.
38. Chekinov and Bogdanov, 'Evoliutsiia sushchnosti i soderzhaniia', p. 39.
39. For example: Jonsson, Oscar and Robert Seely, 'Russian Full-Spectrum Conflict: An Appraisal After Ukraine', *Journal of Slavic Military Studies*, 28, 1 (2015), pp. 1–22; Johnson, 'Russia's Approach to Conflict'; Berzins, 'Russia's New Generation Warfare in Ukraine'; Maigre, 'Nothing New in Hybrid Warfare'; Ruiz Palmer, 'Back to the Future? Russia's Hybrid Warfare'; Galeotti, *Hybrid War or Gibridnaya Voina?*
40. Serebryanikov, Vladimir and Alexander Kapko, 'Nevoyennyye sredstva oboronnoy bezopasnosti Rossii' [The non-military means of the defence security of Russia], *Dialog*, 2 (2000), p. 23.
41. Ibid., pp. 25–6.
42. Ibid., p. 30.
43. Ibid., pp. 29–30.
44. For example: Neklessa, Alexander, 'Upravlyayemyy khaos: dvizheniye k nestandartnoy sisteme mirovykh otnosheniy' [Controlled chaos: a move towards a non-standard system of international relations], *Mirovaia Ekonomika i Mezhdunarodnye Otnosheniia*, 9 (2002), pp. 103–12; Vladimirov, Alexander, 'Strategiya "orgabizovannogo khaosa"' [The strategy of 'organised chaos'], *Prostranstvo i Vremya*, 1 (2010), pp. 53–7; Karadge, Tatyana, 'Teorya khaosa i tekhnologii vneshnego upravleniya' [The chaos theory and the technologies of external control], *Vlast'*, 7 (2015), pp. 41–4; Achkasov, Valery, '"Gibridnyye viyny" kak instrument realizatsii strategii upravlyayemogo khaosa' ['Hybrid wars' as an instrument to implement the strategy of controlled chaos], in Tsygankov, Pavel (ed.), '*Gibridnyye Voyny v khaotiziruyushchemsya mire XXI veka* ['Hybrid wars' in the chaotic world of the twenty-first century], Moscow: Moscow University Press, 2015.
45. For example: Bartosh, Alexander, 'Kompleks podryvnykh tekhnologiy "Tsvetnaya Revolyutsiya: Gibridnaya Voyna" kak ugroza natsional'noy bezopasnosti Rossii' [The complex of 'colour revolution' subversion techniques: 'hybrid war', as a threat to Russia's national security], *Bezopasnost' Yevrazii*, 1 (2015), pp. 245–7; Manoylo, Andrey, 'Tekhnologii "tsvetnykh revolyutsiy" v sovremennykh proyavleniyakh "gibridnykh voyn"' [The techniques of 'colour revolutions' in contemporary expressions of 'hybrid wars], in Tsygankov, *'Gibridnyye Voyny'*; Lobanov, Konstantin, 'Problemy obespecheniya natsional'noy bezopasnosti v protsesse protivodeystviya gibridnym i "tsvetnym" tekhnologiyam' [The problems of providing national secu-

rity in the process of counteracting hybrid and 'colour' techniques], *Srednerusskiy vestnik obshchestvennykh nauk*, 11, 1 (2016), pp. 64–74; Korybko, *Hybrid Wars*.
46. Levada Center, 'Bol'she poloviny rossiyan sozhaleyut o raspade SSSR', 19 April 2016; http://www.levada.ru/2016/04/19/bolshe-poloviny-rossiyan-sozhaleyut-o-raspade-sssr/ (accessed: 5 November 2016).
47. Chekinov and Bogdanov, 'Voennoe iskusstvo', p. 33.
48. See Chapter 6.
49. Freedman, Lawrence, 'Academics and Policy-Making: Rules of Engagement', *Journal of Strategic Studies* (2017), p
50. Ibid.
51. See Chapter 6.
52. Chekinov and Bogdanov, 'Strategicheskoe sderzhivanie', p. 15.
53. Chekinov and Bogdanov, 'Evoliutsiia sushchnosti i soderzhaniia', p. 42.
54. Ibid., p. 43.
55. For example: Jonsson and Seely, 'Russian Full-Spectrum Conflict'; Johnson, 'Russia's Approach to Conflict'; Berzins, 'Russia's New Generation Warfare in Ukraine'; Maigre, 'Nothing New in Hybrid Warfare'; Ruiz Palmer, 'Back to the Future? Russia's Hybrid Warfare'; Galeotti, *Hybrid War or Gibridnaya Voina?*
56. Chekinov and Bogdanov, 'O kharaktere i soderzhanii', p. 24.
57. For example: Jonsson and Seely, 'Russian Full-Spectrum Conflict'; Johnson, 'Russia's Approach to Conflict'; Johnson, 'Russia's Approach to Conflict'; Berzins, 'Russia's New Generation Warfare in Ukrain'; Maigre, 'Nothing New in Hybrid Warfare'; Ruiz Palmer, 'Back to the Future? Russia's Hybrid Warfare'; Galeotti, *Hybrid War or Gibridnaya Voina?*.
58. Chekinov and Bogdanov, 'O kharaktere i soderzhanii', p. 18.
59. Ibid., p. 22.
60. Ibid., pp. 22–3.
61. Ibid., pp. 19–20.
62. Ibid., pp. 22–3.
63. For example: Darczewska, Jolanta, 'The Anatomy of Russian Information Warfare: The Crimean Operation, a Case Study', *Point of View*, 42 (2014); Jonsson and Seely, 'Russian Full-Spectrum Conflict'; Johnson, 'Russia's Approach to Conflict'; Johnson, 'Russia's Approach to Conflict'; Berzins, 'Russia's New Generation Warfare in Ukrain'; Maigre, 'Nothing New in Hybrid Warfare'; Ruiz Palmer, 'Back to the Future? Russia's Hybrid Warfare'; Galeotti, *Hybrid War or Gibridnaya Voina?*
64. Gerasimov, 'Tsennost' nauki v predvidenii'.
65. Gerasimov, Valery, 'Po opytu Sirii: Gibridnaya voyna trebuyet vysokotekhnologichnogo oruzhiya i nauchnogo obosnovaniya' [According to the experience in Syria: hybrid war requires the high-tech weaponry and scientific foundation], *Voyenno-Promyshlennyy Kurier*, 9 (2016); http://vpk-news.ru/articles/29579, (accessed: 28 November 2016).
66. Ibid.

67. Persson, Gudrun (ed.), *Russian Military Capability in a Ten-Year Perspective: 2016*, Stockholm: Swedish Defence Research Agency, 2016, p
68. Ibid.
69. Norberg, Johan, *Training to Fight: Russia's Major Military Exercises 2011–2014*, Stockholm: Swedish Defence Research Agency, 2015, p. 62.
70. Gerasimov, 'Po opytu Sirii'.
71. Persson, *Russian Military Capability in a Ten-Year Perspective*, p. 3.
72. Monaghan, Andrew, 'Russian State Mobilisation: Moving the Country on to a War Footing', Research Paper, Chatham House, London, May 2016, pp. 18–21.
73. Putin, Vladimir, 'Speech at the Meeting of Russian Federation Ambassadors and Permanent Envoys', 30 June 2016, Moscow; http://www.mid.ru/foreign_policy/news/-/asset_publisher/cKNonkJE02Bw/content/id/2338996?p_p_id=101_INSTANCE_cKNonkJE02Bw&_101_INSTANCE_cKNonkJE02Bw_languageId=en_GB (Russian) (accessed: 8 March 2017).
74. 'Sergey Lavrov: Rossiya stolknulas' s bespretsedentnoy informatsionnoy voynoy' [Sergey Lavrov: Russia faces an unprecedented information war], Russia Today, 10 April 2015; https://russian.rt.com/article/84765, (accessed: 8 March 2017).
75. Lavrov, Sergey, 'Speech at the XXII Assembly of the Council of Foreign and Defence Policy', 22 November 2014, Moscow; http://www.mid.ru/vistupleniya_ministra/-/asset_publisher/MCZ7HQuMdqBY/content/id/790194 (Russian) (accessed: 8 March 2017).
76. Putin, Vladimir, 'Speech at an Extended Meeting of the Ministry of Internal Affairs Staff', Moscow, 4 March 2015; http://kremlin.ru/events/president/news/47776 (Russian) (accessed: 8 March 2017).
77. Putin, Vladimir, '2006 Speech at Meeting with the Ambassadors and Permanent Representatives of the Russian Federation', Moscow, 27 June, 2006; http://kremlin.ru/events/president/transcripts/23669 (Russian) (accessed: 8 March 2017).
78. Presidential Decree, 'The Doctrine of the Information Security of the Russian Federation', Moscow, 9 September 2000 (Russian).
79. Presidential Decree, 'The Doctrine of the Information Security of the Russian Federation', Moscow, 5 December 2016 (Russian).
80. Presidential Decree, 'The Military Doctrine of the Russian Federation, Moscow', 21 April 2000 (Russian).
81. Presidential Decree, 'The Military Doctrine of the Russian Federation', Moscow, 5 February 2010 (Russian).
82. Presidential Decree, 'The Military Doctrine of the Russian Federation', Moscow, 25 December 2014 (Russian).
83. Ibid.
84. Presidential Decree, 'The Foreign Policy Concept of the Russian Federation', Moscow, 28 June 2000 (Russian); Presidential Decree, 'The Foreign Policy Concept of the Russian Federation', Moscow, 12 July 2008 (Russian).

85. Presidential Decree, 'The Foreign Policy Concept of the Russian Federation', Moscow, 12 February 2013 (Russian).
86. Presidential Decree, 'The Foreign Policy Concept of the Russian Federation', Moscow, 30 November 2016 (Russian).
87. Presidential Decree no. 683, 'On the Russian Federation National Security Strategy', Moscow, 31 December 2015; http://kremlin.ru/acts/news/51129 (accessed: 5 November 2016).
88. Bogaturov, Alexei, 'Tri pokoleniya vneshnepoliticheskikh doktrin Rossii' [The three generations of Russian foreign policy doctrines], *Mezhdunarodnyye Protsessy*, 5, 1, 3 (2007), pp. 54–69.
89. Putin, Vladimir in Solov'yev, Vladimir, *President*, Masterskaya Movie Company, 2015, (video) (Russian).
90. For example: Panarin, Igor, *Informatsionnyye voyna i kommunikatsii* [Information war and communications], Moscow: Goryachaya Liniya-Telekom, 2015, pp. 200–1.
91. Presidential Decree, 'The Doctrine of the Information Security of the Russian Federation', Moscow, 5 December 2016 (Russian).
92. State Duma, 'Federal Law N 121-FZ: On Amendments to Certain Legislative Acts of the Russian Federation regarding the Regulation of the Activities of Non-profit Organisations That Perform the Functions of a Foreign Agent', Moscow, 20 July 2012, (Russian).
93. State Duma, 'Federal Law N 398-FZ: On Amendments to the Federal Law "On Information, Information Technologies and Protection of Information"', Moscow, 28 December 2013 (Russian); State Duma, 'Federal Law N 97-FZ: On Amendments to the Federal Law "On Information, Information Technologies and Protection of Information" and Certain Legislative Acts of the Russian Federation on the Regulation of the Exchange of Information Using Information–Telecommunication Networks', Moscow, 21 July 2014 (Russian); State Duma, 'Federal Law 374-FZ: On Amendments to the Federal Law "On Combating Terrorism" and Certain Legislative Acts of the Russian Federation to Establish Additional Measures to Counter Terrorism and Ensure Public Safety', Moscow, 6 July 2016 (Russian).
94. State Duma, 'Federal Law N 305-FZ: On Amendments to the Law of the Russian Federation "On Mass Media"', Moscow, 14 October 2014 (Russian).
95. For example: Gaynullina, Alina, 'Internet-kompanii kritikuyut antipiratskiy zakonoproyekt deputatov GD' [The internet companies criticise the anti-piracy legislation of the State Duma], Ria-Novosti, 13 June 2013; https://ria.ru/society/20130613/943211875.html (accessed: 13 March 2017); Surnacheva, Elizaveta, 'Ogranichennyye Dumoy' [Restricted by the State Duma], Kommersant.ru, 29 September 2014; http://www.kommersant.ru/doc/2573623 (accessed: 13 March 2017); 'Russian MPs Back Law on Internet Data Storage', BBC News, 5 July 2014; http://www.bbc.com/news/world-europe-28173513 (accessed:

13 March 2017); 'Russia: Writers and Academics Speak Out against Law on Foreign Agents', PEN American Center, 5 February 2016; https://pen.org/blog/russia-writers-and-academics-speak-out-against-law-foreign-agents (accessed: 13 March 2017); Dyomkin, Denis, 'Council of Europe Tells Putin of Concern over Russian NGO Law', Reuters, 20 May 2013; http://www.reuters.com/article/us-russia-europe-ngos-idUSBRE94J0S120130520 (accessed: 13 March 2017).

96. Kiselev, Valery and Ivan Vorob'ev, 'Gibridnyye operatsii kak novyy vid voyennogo protivoborstva [Hybrid Operations as a New Type of Confrontation]', *Voennaya Mysl'*, 5 (2015), p. 41.
97. Lavrov, 'Speech at the XXII Assembly'.
98. Zyuganov, Gennady, speech during a meeting of President Putin with the leaders of the Russian parliamentary groups. See 'Vstrecha s rukovoditelyami parlamentzkikh fraktzii' [A meeting with the leaders of the parliamentary groups], Moscow, 14 July 2016; http://kremlin.ru/events/president/news/52513 (accessed: 16 March 2017).
99. Levada Center, 'Reaktsiya Zapada na politiku Rossii: kritika, vrazhdebnost', sanktsii' [The West's reaction to Russian policy: critique, hostility, sanctions], 2 December 2015; http://www.levada.ru/2015/11/02/reaktsiya-zapada-na-politiku-rossii-kritika-vrazhdebnost-sanktsii/ (accessed: 16 March 2017).
100. Emerson, Michael, 'Putin: The Morally Depraved Degradation of a Strong State Leader', *CEPS European Neighbourhood Watch*, 114 (April 2015), p. 2.

CONCLUSIONS: THE RISE OF RUSSIAN 'HYBRID WAR'—LESSONS FOR THE WEST

1. For example: Connell, Michael and Sarah Vogler, 'Russia's Approach to Cyber Warfare', Occasional Paper, Center for Naval Analysis, Arlington, September 2016; Nimmo, Ben, 'Anatomy of an Info-War: How Russia's Propaganda machine Works, and How to Counter It', Central European Policy Institute (CEPI), Bratislava, 2015; Giles, Keir, 'The Next Phase of Russian Information Warfare', in Renz and Smith, 'Russia and Hybrid Warfare'; NATO Strategic Communications Centre of Excellence, 'Internet Trolling as a Tool of Hybrid Warfare: The Case of Latvia', Riga, 2016; NATO Strategic Communications Centre of Excellence, Analysis of Russia's Information Campaign against Ukraine, Riga, 2015; Intelligence Community Assessment, 'Assessing Russian Activities and Intentions in Recent US Elections', Washington, DC, 6 January 2017; European Union Institute for Security Studies (EUISS), 'EU Strategic Communications with a View to Counteracting Propaganda, Paris', 19 May 2016.
2. Gray, Colin, *Perspectives on Strategy*, Oxford: Oxford University Press, 2013, p. 2.
3. Hoffman, Frank, *Conflict in the 21st Century: The Rise of Hybrid Warfare*, Arlington: Potomac Institute for Policy Studies, 2007, p. 29.

4. Hoffman, Frank, 'Hybrid vs. Compound Wars', *Armed Forces Journal* (October 2009), p. 2.
5. Mihara, Robert, 'Beyond Future Threats: A Business Alternative to Threat-Based Strategic Planning', *Infinity Journal*, 2, 3 (Summer 2012), p. 25.
6. Dan, Thomas Bruscino and Alex Ryan, 'Why Hybrid Warfare is Tactics Not Strategy: A Rejoinder to "Future Threats and Strategic Thinking"', *Infinity Journal*, 2, 2 (Spring 2012), p. 25.
7. Joint Irregular Warfare Center, 'Irregular Adversaries and Hybrid Threats, an Assessment: 2011', Norfolk, 2011, p. 24.
8. Breedlove, Philip, 'Foreword', in Lasconjarias, Guillaume and Jeffrey Larsen (eds), *NATO's Response to Hybrid Threats*, (Rome: NATO Defence College, 2015, p. xxii.
9. Ruiz Palmer, Diego, 'Back to the Future?' Russia's Hybrid Warfare, Revolutions in Military Affairs, and Cold War Comparisons', Research Paper no. 120, NATO Defence College, Rome, October 2015, p. 2.
10. Tsygankov, Pavel, 'Gibridnyye Voyny: ponyatiya, interpretatsii i real'nost' [Hybrid wars: definitions, interpretations and reality]', in Tsygankov, Pavel (ed.), *'Gibridnyye Voyny v khaotiziruyushchemsya mire XXI veka* ['Hybrid wars' in the chaotic world of the twenty-first century], Moscow: Moscow University Press, 2015, pp. 5–6.
11. Gerasimov, Valery, 'Po opytu Sirii: Gibridnaya voyna trebuyet vysokotekhnologichnogo oruzhiya i nauchnogo obosnovaniya' [According to the experience in Syria: hybrid war requires the high-tech weaponry and scientific foundation], *Voyenno-Promyshlennyy Kurier*, 9 (2016); http://vpk-news.ru/articles/29579, (accessed: 28 November 2016).
12. Longman Dictionary of Contemporary English, 6th edn, Harlow: Pearson Education, 2014, p. 905.
13. 'Hybrid War: Does It Even Exist?', *NATO Review*, 7 May 2015; http://www.nato.int/docu/review/2015/Also-in-2015/hybrid-modern-future-warfare-russia-ukraine/EN/ (accessed: 21 March 2017).
14. Smith, Paul, *On Political War*, Washington, DC: National Defense University Press, 1989.
15. Mearsheimer, John, 'Why the Ukraine Crisis Is the West's Fault: The Liberal Delusions That Provoked Putin', *Foreign Affairs*, 93 (2014), pp. 77–89.
16. Von Clausewitz, Carl, On War, ed. and trans. Michael Howard and Peter Paret, Oxford: Oxford University Press, 2008, p. 40.
17. Ibid., p. 28.
18. Ibid., p. 13.
19. Ibid., p. 255.
20. Renz, Bettina and Hanna Smith, 'PART 2: A Dangerous Misuse of the Word "War"? "Hybrid Warfare" as a Quasi-theory of Russian Foreign Policy', in Renz, Bettina and Hanna Smith (eds), 'Russia and Hybrid Warfare: Going Beyond the Label', *Papers Aleksanteri*, 1 (2016), p. 11.
21. Samuel Charap, 'The Ghost of Hybrid War', *Survival*, 57, 6 (2015/16), p. 52.

22. For example: Gol'ev, A., 'Voyna kak faktor sovremennogo politicheskogo protsessa' [War as a factor in the contemporary political process], *Vestnik MGLU*, 24, 684 (2013), pp. 113–23.
23. Bousquet, Antoine, 'The Concept of War', in Berenskoetter, Felix (ed.), *Concepts in World Politics*, London: Sage Publications, 2016.
24. Gareev, Makhmut, 'Voyna i voennaya nauka na sovremennom etape' [Contemporary war and military science], *Voenno-Promyshlennyy Kur'er*, 13 (2013); http://vpknews.ru/sites/default/files/pdf/VPK_13_481.pdf (accessed: 1 March 2017).
25. Gareev, Makhmut, 'Struktura I osnovnoe soderzhanie novoy voennoy doktriny' [The structure and content of the new military doctrine], *Voenno-Promyshlennyy Kur'er*, 3 (2007), http://www.vpk-news.ru/sites/default/files/pdf/issue_169.pdf (accessed: 1 March 2017).
26. Chekinov, Sergey and Sergey Bogdanov, 'Evoliutsiia sushchnosti i soderzhaniia poniatiia "voi'na" v XXI stoletii' [The evolution of the nature and content of the Concept of 'war' in the twenty-first century], *Voennaya Mysl'*, 1 (2017), p. 39.
27. Qiao, Liang and Xiangsui Wang, *Unrestricted Warfare: China's Master Plan to Destroy America*, Panama City: Pan American Publishing Company, 2002, p. 118.
28. Smith, *On Political War*, p. 3.
29. NATO Strategic Communications Centre of Excellence, 'Analysis of Russia's Information Campaign', p. 26.
30. Freedman, Lawrence, 'Stop Overestimating the Threat Posed by Russia's "New" Form of Warfare', World Economic Forum, 4 January 2017; https://www.weforum.org/agenda/2017/01/stop-overestimating-the-threatposed-by-russia-s-new-form-of-warfare/ (accessed: 22 February 2017).
31. For example: Volkovsky, Nikolay, *Istroiya informatzionnykh voyn* [The history of information wars], vol. 2, Saint Petersburg: Poligon, 2013; Smith, *On Political War*.
32. Freedman, 'Stop Overestimating the Threat'.
33. Ruiz Palmer, 'Back to the Future? Russia's Hybrid Warfare', p. 2.
34. For example: Winter, Charlie, *Documenting the Virtual 'Caliphate'*, London: Quilliam, 2015; Fernandez, Alberto, *Here to Stay and Growing: Combating ISIS Propaganda Networks*, Washington, DC: Brookings Institution, 2015); Stern, Jessica and J.M. Berger, *ISIS: The State of Terror*, New York: HarperCollins, 2015; Atwan, Abdel Bari, *Islamic State: The Digital Caliphate*, Oakland: University of California Press, 2015.
35. Freedman, Lawrence, 'Strategic Studies and the Problem of Power', in Mahnken, Thomas and Joseph Maiolo (eds), *Strategic Studies: A Reader*, Abingdon: Routledge, 2008, p. 32.
36. Qiao and Wang, *Unrestricted Warfare*, p. 117.
37. Heuser, Beatrice, *The Evolution of Strategy: Thinking War from Antiquity to the Present*, Cambridge: Cambridge University Press, 2010, p. 3.

38. For example: Mihara, Robert, 'Beyond Future Threats: A Business Alternative to Threat-Based Strategic Planning', *Infinity Journal*, 2, 3 (Summer 2012); Friedman, Brett, 'Blurred Lines: The Myth of Guerrilla Tactics', *Infinity Journal*, 3, 4 (Winter 2014); Cox, Bruscino and Ryan, 'Why Hybrid Warfare Is Tactics Not Strategy'.
39. Huber, Thomas (ed.), *Compound Warfare: That Fatal Knot*, Fort Leavenworth: Combat Studies Institute, US Army Command and General Staff College, 2002; Murray, Williamson and Peter Mansoor (eds), *Hybrid Warfare: Fighting Complex Opponents from the Ancient World to the Present*, Cambridge: Cambridge University Press, 2012.
40. Freedman, 'Stop Overestimating the Threat'.
41. Dull, Jonathan, *A Diplomatic History of the American Revolution*, New Haven: Yale University Press, 1985.
42. Katkov, George, 'German Foreign Office Documents on Financial Support to the Bolsheviks in 1917', *International Affairs*, 32, 2 (1956), pp. 181–9; Kerensky, A. and George Katkov, 'German Foreign Office Documents on Financial Support to the Bolsheviks in 1917', *International Affairs*, 32, 4 (1956), pp. 534–40.
43. Lawrence, T.E., *Seven Pillars of Wisdom*, Ware: Wordsworth Editions, 1997; Murphy, David, *The Arab Revolt 1916–18, Lawrence Sets Arabia Ablaze*, London: Osprey, 2008.
44. Cantrell, Robert, *Understanding Sun Tzu on the Art of War*, Arlington: Center for Advantage, 2003, p. 24.
45. Smith, *On Political War*, p. 3.
46. Bartles, Charles, 'Getting Gerasimov Right', *Military Review*, 96 (2016), p. 31.
47. Presidential Decree, 'The Military Doctrine of the Russian Federation', Moscow, 25 December 2014 (Russian).
48. Ibid.
49. Galeotti, Mark, *Hybrid War or Gibridnaya Voina? Getting Russia's Non-Linear Military Challenge Right*, Prague: Mayak Intelligence, 2016, p. 25.
50. Presidential Decree, 'The Military Doctrine of the Russian Federation', Moscow, 25 December 2014 (Russian).
51. Ibid.
52. Freedman, 'Stop Overestimating the Threat'.
53. Tyutchev, Fyodor, *Izbrannoe* [Selected], (Rostov-on-Don: Phenix, 1996).
54. Churchill, Winston, 'The Russian Enigma', BBC Broadcast, London, 1 October 1939; http://www.churchill-society-london.org.uk/RusnEnig.html (accessed: 26 March 2015).
55. See Chapter 7.
56. Kennan, George, 'Telegram: The Charge in the Soviet Union (Kennan) to the Secretary of State', Moscow, 22 February 1946; http://nsarchive.gwu.edu/coldwar/documents/episode-1/kennan.htm (accessed: 26 March 2017).
57. Association for Slavic, East European, and Eurasian Studies (ASEEES), 'The State

of Russian Studies in the United States', University of Pittsburgh, Pittsburgh, July 2015, p. 4.
58. Ibid.
59. Rojansky, Matthew, 'George F. Kennan, Containment, and the West's Current Russia Problem', Research Paper no. 127, NATO Defence College, Rome, January 2015, p. 9.
60. Association for Slavic, East European, and Eurasian Studies (ASEEES), 'State of Russian Studies', p. 45.
61. Rojansky, 'George F. Kennan', p. 10.
62. Ibid.
63. Hill, Fiona quoted in Demirjian, Karoun, 'Lack of Russia Experts Has Some in U.S. Worried', *Washington Post*, 30 December 2015; https://www.washingtonpost.com/news/powerpost/wp/2015/12/30/lack-of-russia-experts-has-the-u-s-playing-catch-up/?utm_term=.45b9a6ce3bbb (accessed: 23 March 2017).
64. Kennan, 'Telegram'.
65. See Chapter 5.
66. Smith, *On Political War*, Chapter 9.
67. Ibid.
68. Ibid.
69. Kennan, 'Telegram'.
70. Smith, *On Political War*, pp. 203–4.
71. Cantrell, *Understanding Sun Tzu*, p. 91.
72. Galeotti, *Hybrid War or Gibridnaya Voina?*, pp. 91–2.
73. Radin, Andrew, *Hybrid Warfare in the Baltics: Threats and Potential Responses*, Santa Monica: Rand Corporation, 2017.
74. Cuddington, Danielle, 'Support for NATO Is Widespread among Member Nations', PewResearchCenter, 6 July 2016; http://www.pewresearch.org/fact-tank/2016/07/06/support-for-nato-is-widespread-among-member-nations (accessed: 28 March 2017).
75. Giles, Keir, 'Russia's "New" Tools for Confronting the West: Continuity and Innovation in Moscow's Exercise of Power', Research Paper, Chatham House, London, March 2016, p. 59.
76. Kiley, Jocelyn, 'U.S. Public Sees Russian Role in Campaign Hacking, But Is Divided Over New Sanctions', PewResearchCenter, 10 January 2017, http://www.pewresearch.org/fact-tank/2017/01/10/u-s-public-says-russia-hacked-campaign (accessed: 28 March 2017).
77. Sherr, James, 'The New East-West Discord Russian Objectives, Western Interests', Clingendael Report, Netherlands Institute of International Relations, The Hague, December 2015, p. 74.
78. Clausewitz, *On War*, p. 160.
79. Kennan, 'Telegram'.

80. Rojansky, 'George F. Kennan', p. 10.
81. Charap, Samuel and Jeremy Shapiro, 'Consequences of a New Cold War', *Survival*, 57, 2 (2015), pp. 37–46.
82. Ibid.
83. McFaul, Michael quoted in Demirjian, 'Lack of Russia Experts'.
84. Galeotti, *Hybrid War or Gibridnaya Voina?*, pp. 98–9.
85. Dzyaloshinskiy, Iosif, 'Kul'tura, zhurnalistika, tolerantnost' [Culture, journalism, tolerance], in Dzyaloshinskiy, Iosif and Marina Dzyaloshinskaya (eds), *Rossiyskiye SMI: kak sozdayetsya obraz vraga* [Russian mass media: how the image of an enemy is created], Moscow: Moskovskoe Byuro po Pravam Cheloveka 'Academia', 2008, pp. 24–5.
86. Sherr, 'New East–West Discord', p. 73.
87. Mearsheimer, 'Why the Ukraine Crisis Is the West's Fault', p. 82.

P.S. A LESSON FOR RUSSIA

1. Presidential Decree, 'The Foreign Policy Concept of the Russian Federation', Moscow, 30 November 2016, (Russian).
2. Von Clausewitz, Carl, On War, ed. and trans. Michael Howard and Peter Paret, Oxford: Oxford University Press, 2008, p. 214.
3. Ibid., p. 213.
4. Ibid., p. 214.
5. McFaul, Michael quoted in Demirjian, Karoun, 'Lack of Russia Experts Has Some in U.S. Worried', *Washington Post*, 30 December 2015; https://www.washingtonpost.com/news/powerpost/wp/2015/12/30/lack-of-russia-experts-has-the-u-splaying-catch-up/?utm_term=.45b9a6ce3bbb (accessed: 23 March 2017).

INDEX

Academy of Geopolitical Problems, 95
'active measures' (*aktivinye meropriatia*), 3–4, 169
Adamsky, Dima, 113, 114, 124
Afghanistan
    Soviet War (1979–89), 27, 34
    Taliban, 80
    US-led war (2001–14), 112, 119, 160
aggressive-diplomacy, 64–6
Alexander the Great, 19
Algeria, 65
Almaz-Antey, 98
American Civil War (1861–5), 21
American Revolutionary War (1775–83), 27, 65, 162
al-Aqsa Intifada (2000–2005), 23
Arab Revolt (1916–18), 162
Arab Spring (2011), 70
Arab-Israeli conflict, 63, 70
Argentina, 52
Armed Forces Journal International, 37
Arquilla, John, 31, 32
Association for Slavic, East European, and Eurasian Studies (ASEEES), 167
asymmetry, 18, 25, 40, 135, 160
Atlantic Council, 110

Australia, 31, 32, 63

Balkans, conflict in (1991–2001), 34, 70, 129, 134
ballistic missiles, 55
Baltic states, 111, 114, 120–21, 158, 171
Barons' War, First (1215–17), 65
Battle of Bint-Jbeil (2006), 45, 154
Battle of Grozny (1994–5), 45, 154
Battle of Mogadishu (1993), 45, 154
Battle of Sedan
    Franco-Prussian War (1870), 70
    Second World War (1940), 4
Battle of the Somme (1916), 4
Battle of Verdun (1916), 23
Battle of Waterloo (1815), 23
Behnke, Andreas, 119
Belgrade, Serbia, 51–2
Bin Laden, Osama, 16, 80, 81
Bint-Jbeil, battle of (2006), 45, 154
Blank, Stephen, 31
blitzkrieg, 165
Bogdanov, Sergey, 113, 127–44, 150, 158, 159, 163
Bolsheviks, 39, 51, 55, 162; *see also* Russian Revolution
Bond, Margaret, 38

INDEX

Boot, Max, 20, 55, 102
Breedlove, Philip Mark, 106, 110
Brexit referendum (2016), 122
Brezhnev, Leonid, 54
British Empire, 76, 87, 90, 162
Brookings Institution, 168
Bucharest, Romania, 112
Buenos Aires, Argentina, 52

'call-for-action' narrative, 113
Carmichael, Stokely, 69
catastrophic challenges, 28, 33
Catherine II 'the Great', Empress and Autocrat of All the Russias, 49
Cebrowski, Arthur K., 77
Centre for Emerging Threats and Opportunities, 37
Centre for Military and Strategic Studies, 127
Chatham House, 116, 175
Chechnya, 34, 45, 83, 112, 115, 154
Chekinov, Sergey, 113, 127–44, 150, 158, 159, 163
China, 53, 54
  Civil War (1927–49), 23, 66, 68
  Cultural Revolution (1966–76), 54
  Indochina War, First (1946–54), 65
  Taiwan Strait Crisis (1954–5), 65
  Unrestricted Warfare (Qiao and Wang), 12–19, 32, 34, 66, 159, 161
Christianity, 53
  Orthodoxy, 49, 76
  Protestantism, 76
Churchill, Winston, 66, 166
von Clausewitz, Carl, 6, 42, 64, 66, 109, 128, 135, 157, 177, 178
Clinton, Hillary, 122, 123
cocktails, 14, 19, 32, 36, 159, 161–3
Cohn-Bendit, Daniel, 69
Cold War, 5, 6

'active measures' (*aktivinye meropriatia*), 3–4, 169
and *gibridnaya voyna*, 97, 133, 137, 138, 139, 149, 157, 165, 169, 170
information war, 87, 89, 138, 171–2
Long Telegram (1946), 166, 168–9, 170, 173
network-centric warfare, 82, 138
nostalgia, 124
political warfare, 96–7, 159, 165, 170, 172–3
repetition of, 3, 124, 149, 165, 173–5
subversion-war, 49, 52, 54, 62–4, 65, 72, 138
and travel, restrictions on, 167
collapse of Soviet Union (1991), 72, 91, 97, 149, 174
and *gibridnaya voyna*, 97, 133, 137, 138, 139, 149, 157, 165, 169, 170
and Marxism–Leninism, 71, 148, 170
military philosophy, 71, 91, 97, 165
and information war, 87–8, 89
NATO, effect on, 119
and net-centric war, 82
public's views on, 139, 152
and subversion-war, 54, 72
Colour Revolutions, 93, 133–4, 139
Command and Control Research Program (CCRP), 77
communications technology, 44, 130, 164
Communism, 39, 49, 52, 53, 55, 56
Communist Manifesto (Marx and Engels), 55
compound warfare, 12, 24–7, 31, 32–3, 36, 154, 158, 159
Conflict Studies Research Centre (CSRC), 116
continent-oriented civilisations, 87

226

INDEX

controlled chaos, 139
cosmopolitanism, 56
counterinsurgency, 34
Cox, Dan, 102
Crimea annexation (2014), 175
   and conceptual puzzle, 119, 121
   and force multipliers, 117
   as hybrid warfare, 101, 106, 108–9, 112–14, 115, 117, 119–23, 151, 154, 156–7
   as international relations problem, 124, 125
   and information warfare, 144
   NATO reaction to, 107, 108–9, 119
   politicisation, 119, 120, 121, 123, 125
   and Russian studies, decline in, 168
Crimean War (1853–6), 60
criminality, 11, 15, 32, 33, 154
Cultural Revolution (1966–76), 54
cyber warfare, 2, 132, 153
   in Chechnya, 115
   in Georgia, 115
   and *gibridnaya voyna*, 135, 137, 139, 156, 158, 159
   Hoffman's hybrid warfare, 32
   and media, 172
   National Defence Strategy, US (2005), 28
   and NATO, 119, 120, 121, 124
   politicisation of, 119–24, 139
   Swedish National Defence College, 105
   in Ukraine, 106, 108, 111, 115–18, 123
   in United States, 118
   unrestricted warfare, 15
   and war, misuse of term, 157
Czech Republic, 116
Czechoslovakia (1918–92), 137

Delbrück, Hans, 66

democracy, 53, 56, 118, 120, 121, 122, 123, 124–5, 175
Democratic Party, US, 118, 122
Department of Defense, US, 27–9, 45
dimensions of war, 59
disruptive challenges, 28
Dugin, Aleksandr Gelyevich, 7, 75–84
   and adversary, nature of, 78, 79
   and agents of influence, 80–81
   and artificial networks, 80
   and control, understanding of, 78–9
   and *gibridnaya voyna*, 92, 95, 96, 117, 128, 133, 134, 136, 137, 158
   and natural networks, 80
   and Panarin, 89, 90
   Putin, support for, 89
Dutschke, Rudi, 69

Eastern Europe
   'call-for-action' narrative, 113
   NATO deployment in, 3, 114, 158
   politicisation of discourse, 119, 120, 124
   Russian studies, 66, 110–11, 119, 168, 172–3
   Russian threat to, 171, 172–3
Echevarria, Antulio, 24, 103
economic interdependence, 130, 134, 147–8
economic warfare, 7, 90, 92
   and Chekinov and Bogdanov, 130, 131, 132, 134, 135, 137, 138, 139, 140, 142
   in First World War (1914–18), 162
   and fourth generation warfare, 19
   and Gerasimov, 143
   and hybrid warfare/*gibridnaya voyna*, 37, 93–8, 104, 105, 150, 155, 156, 162, 163, 169
   and information warfare, 85, 88, 89, 90

INDEX

and net-centric warfare, 81, 83
against Russian Federation, 145, 146, 147, 148
against Soviet Union, 137, 139, 159
and subversion-war, 60, 66, 70, 72
against Ukraine, 106, 108, 109, 111, 154
and unrestricted warfare, 15, 16, 17, 18
against West, 170–71
Edison, Thomas, 55
Egypt, 65
Einstein, Albert, 64
'enemies of the people' (*vragi naroda*), 3
Engels, Friedrich, 66
England, 65; *see also* United Kingdom
Estonia, 113, 120–21
Eurasianism, 76, 84, 89
European Union (EU), 122, 125
Evans, Michael, 31

Fabianism, 53
feminisation, 56
fifth column, 90, 134
financial interdependence, 130, 134, 147–8
financial warfare
and cyber warfare, 32
and hybrid warfare/*gibridnaya voyna*, 1, 7, 90, 93, 94, 104, 109, 135, 145, 150, 170
and information warfare, 85, 88, 89
and net-centric warfare, 79
Soviet use of, 170
and subversion-war, 62
United States use of, 145
and unrestricted warfare, 15
Finland, 112
First Barons' War (1215–17), 65
first generation warfare, 19, 20

First World War (1914–18), 3, 4, 21–2, 23, 55, 60
Arab Revolt (1916–18), 162
France, 22
Germany, 21–2, 58, 162, 165
Russia, 39, 51, 162
second generation warfare, 21–2
Treaty of Brest-Litovsk (1918), 62
United Kingdom, 162
Foreign Policy Concept, Russian Federation, 147–8
Forrestal, James, 125
Foucault, Michel, 3
fourth generation warfare (4GW), 12, 19–24, 57, 66, 154, 158, 159
dispersion on battlefield, 21–3, 33, 59
hybrid warfare, 31, 32, 33, 34, 36, 37, 38, 44
insurgency, focus on, 23, 24, 36
and judo, 56
media manipulation, 62
Peace of Westphalia (1648), 19, 57
political will to fight, 19, 23–4, 33, 57, 62
France
Algerian War (1954–62), 65
American Revolutionary War (1778–83), 65, 162
Barons' War, First (1215–17), 65
Indochina War, First (1946–54), 65, 66
Indochina War, Second (1955–75), 27
May 1968 protests, 69
Napoleonic Wars (1803–15), 20, 21, 24, 26, 70, 151
Prussian War (1870–71), 70
Revolution (1789–99), 53, 55
second generation warfare, 22
Suez Crisis (1956), 65

INDEX

Freedman, Lawrence, 104, 112, 114, 122, 139–40, 160, 165
Friedman, Brett, 103

Galeotti, Mark, 116–18, 164, 171
Gareev, Makhmut, 92, 132, 158
Gaza War (2008–9), 38
Georgia, 115
Gerasimov, Valery, 96, 98, 113, 115, 128, 141, 142–4, 150, 163
Germany
  blitzkrieg, 165
  Bolsheviks, funding of, 162
  federal election (2017), 122
  First World War (1914–18), 21–2, 58, 162, 165
  Franco-Prussian War (1870–71), 70
  Movement 2 June, 69
  Munich Security Conference, 145
  Second World War (1939–45), 22, 40, 52, 70, 151, 165
  third generation warfare, 22
Giáp, Võ Nguyên, 66
Giles, Keir, 116–18, 171
Global War on Terror, 119
globalisation, 14, 15, 16, 17, 19, 23, 81, 130, 134, 137
Golovin, Nikolai, 52
Gray, Colin, 31
Grigor'yev, Yuri, 85–6
Grozny, battle of (1994–5), 45, 154
Grybauskaitė, Dalia, 121
Guevara, Ernesto 'Che', 66
Gulf War (1990–91), 12, 16–17, 129
Gunneriusson, Håkan, 105
Gunpowder Revolution, 55

half-war, 64–5
Hammes, Thomas, 20, 23, 32, 57, 59
Hezbollah, 11, 34–6, 37, 38, 39–40, 45, 112

Higher Military Science Courses, 52
Hill, Fiona, 168
Hitler, Adolf, 151
Hoffman, Bruce, 31
Hoffman, Frank, 7, 11, 31–8, 40, 43–5, 105, 154, 156, 158, 159, 161
  at Centre for Emerging Threats and Opportunities, 37
  criticisms of, 102–3
  and fourth generation warfare (4GW), 31, 32, 33, 38
  and *gibridnaya voyna*, 91, 92, 93, 94, 95, 96, 155, 158
  influences, 31–2
  and Lebanon War (2006), 11, 34–6, 37, 102
  on National Defence Strategy (2005), 27, 33
  and NATO, 101, 104, 110
  novelty of hybrid warfare, 44, 161
  and Ukrainian Crisis (2014), 109
  Unrestricted Warfare (Qiao and Wang), 13
  and US military culture, 43
House of Representatives, US, 38
Huber, Thomas, 24–7, 159, 161
human rights, 56, 80
humanism, 53
Huntington, Samuel, 19

ideology, 7, 35, 39–40
Imperial Nicholas Military Academy, St Petersburg, 51, 93
Imperial Novorossiya University, Odessa, 49
Indochina Wars
  First (1946–54), 65, 66
  Second (1955–75), 23, 27, 40, 63, 69–70, 112
Industrial Revolution, 55
Information Security Doctrines, Russian Federation, 146–7

INDEX

information technology, 14–15
information war, 2, 7, 72–3, 75, 84–90, 132, 155, 156, 157, 158, 159, 165
   Chekinov and Bogdanov, 129–30, 135, 140, 141, 142
   Gerasimov, 143
   and *gibridnaya voyna*, 92, 94–7, 127, 128, 133, 134, 135, 136, 137, 158, 163
   Hezbollah's use of, 35
   history of, 130
   and hybrid warfare, 35, 38, 94
   meaningful system of management, 88–9
   political elite, 88
   private-governmental partnership, 89
   and Russian Federation's use of, 2, 108, 115–18, 139, 144–5, 153, 160
   social objects, 86–7
   and Soviet Union, 87–8, 89, 117, 170, 171–2
   and Ukrainian Crisis (2014), 108, 115–18, 139, 144–5, 160
   and unrestricted warfare, 15
   Western use of, 133, 137, 139, 145–52, 153
   see also Panarin, Igor Nikolaevich
Institute for the Research of War and Peace, 52
Institute of International Relations, Prague, 116
insurgency, 19, 23, 24, 33, 34, 36, 66–8, 70
International Institute for Strategic Studies (IISS), 106
International Monetary Fund (IMF), 81
international organisations, 16, 23
Internet, 44, 164

Iran, 36, 39, 80
Iraq
   Gulf War (1990–91), 12, 16–17, 129
   Saddam era (1979–2003), 80
   US-led war (2003–11), 38, 40–41, 112, 160
Irish War of Independence (1919–21), 34
irregular challenges, 28
ISIS (Islamic State of Iraq and al-Sham), 160
Israel, 63, 70
   al-Aqsa Intifada (2000–2005), 23
   Gaza War (2008–9), 38
   Lebanon War, Second (2006), 11, 34–6, 37, 38, 39–40, 45, 102, 112, 154

Johnson, Lyndon Baines, 69
Johnson, Richard, 38–41
Joint Chiefs of Staff, 12
Jomini, Antoine-Henri, 66
Jordan, Larry, 38
judo, 56

Kalinina, Lyudmila Emmanuilovna, 51
Kalnin, Immanuel, 51
Kalugin, Oleg, 3–4
Kapko, Alexander, 137, 138
Kautsky, Karl, 66
Kennan Institute, 167
Kennan, George, 87, 166, 168–9, 170, 173–4
KGB (Komitet gosudarstvennoy bezopasnosti), 3, 88, 148
Kherson, Ukraine, 49
Khrushchev, Nikita, 54
Kilcullen, David, 104
King, Paul, 109
Kissinger, Henry, 135

INDEX

Knox, MacGregor, 55
Kofman, Michael, 114, 120, 125
Kokoshin, Andrey, 130
Kosovo War (1998–9), 129
Kutuzov, Mikhail, 64

Lafayette, Marquis de, *see* Motier, Gilbert
Larsen, Jeffrey, 1
Lasconjarias, Guillaume, 1
Lasica, Daniel, 38
Latin America, 95
Latvia, 116, 120–21
Lavrov, Sergey, 145, 151
Lebanon, 11, 34–6, 37, 38, 39–40, 45, 102, 112, 154
Leer, Genrikh Antonovich, 66
leftism, 53
Lenin, Vladimir, 54
Levada Centre, 152
liberalism, 53, 56, 83
Lichtenstein, 52
Liddell Hart, Basil, 135
Lind, William, 19, 20, 21, 22, 23, 56–7, 59, 159
Lisichkin, Vladimir, 86
Lithuania, 121
Lomonosov Moscow State University, 1, 76, 91, 94
Long Telegram (1946), 166, 168–9, 170
Louis VIII, King of France, 65
Ludendorff, Erich, 66
Lukashevich, Sergey, 51

Mansoor, Peter, 102, 161
Mao Zedong, 54, 66, 68, 135
Marcuse, Herbert, 54
Marine Corps, US, 19, 22, 32, 37, 102
Marshall Plan, 174
Marx, Karl, 66

Marxism–Leninism, 71, 148, 170
McCulloh, Timothy, 38–41
McFaul, Michael, 174
McWilliams, Sean, 38
Mearsheimer, John, 175
Merkel, Angela, 122
Messner, Alexander Yakovlevich, 50
Messner, Evgeny Eduardovich, 5, 7, 49–73, 75
  aggressive-diplomacy, 64–6
  and Cold War, 49, 52, 54, 62–4, 65, 72, 138
  conservatism, 53, 54, 56, 57, 67
  half-war, 64–5
  and *gibridnaya voyna*, 92, 95, 96, 117, 128, 133, 134, 136, 137, 158
  nationalisation of war, 57–9, 60, 65, 67
  psychological dimension, 59–62, 65, 67–8
  translation of, 66
  'World Revolution' (*Vsemirnaya Revolutsiya*), 55–7, 59, 60, 67
Mihara, Robert, 103
Mikhailovsky Artillery School, St Petersburg, 50
Mikhnevich, Nikolai, 66
Military Doctrine, Russian Federation, 147, 163
Military Revolution, 20, 25
Mogadishu, battle of (1993), 45, 154
von Moltke, Helmuth von, 70
Monaghan, Andrew, 144
Moorer, Thomas, 12
Morozov, Evgeny, 54
du Motier, Gilbert, Marquis de Lafayette, 65
Mujahedeen, 27, 34
Munich Security Conference, 145
Murray, Williamson, 20, 55, 102, 161
mutiny-war, 66

INDEX

Napoleon I, Emperor of the French, 66, 151
'Napoleon in Spain and Naples' (Huber), 24
Napoleonic Wars (1803–15), 20, 21, 24, 26, 70, 151
National Defence Strategy, US (2005), 12, 27–9, 33, 154
nationalisation of war, 57–9, 60, 65, 67
nationalism, 20, 21
Nazi Germany (1933–45), 40, 52, 70, 151
net-centric war, 7, 72, 75–84
    adversary, nature of, 78, 79
    agents of influence, 80–81
    artificial networks, 80
    control, understanding of, 78–9
    and *gibridnaya voyna*, 92, 95–7, 117, 127, 128, 132, 133, 134, 136, 137, 155, 158, 159
    natural networks, 80
    and Soviet Union, collapse of (1991), 82
network-centric warfare, 76–7, 79, 81–4
New Zealand, 63
new-generation warfare, 131, 136, 137, 140–44, 150, 151, 158, 159, 163, 164
nihilism, 53, 54
non-governmental organisations (NGOs), 16, 23, 80, 83, 149
Norberg, Johan, 144
North Atlantic Treaty Organization (NATO), 1, 3, 104–25, 138, 140, 154, 155, 157, 164
    Advance Research Workshop, 112
    Afghanistan War (2001–14), 119
    Allied Command Transformation (ACT), 104
    Bi-Strategic Command Capstone Concept, 104
    budget, 3
    'Countering Hybrid Threats' conference (2015), 112
    Defence College (NDC), 1, 106, 107, 111–12
    Eastern Europe, deployment in, 3
    Excellence Centres, 105, 116, 120
    Global War on Terror, 119
    and hybrid warfare, 101, 104–25, 154, 157, 159
    identity crisis, 119–20, 124
    Readiness Action Plan (RAP), 114
    'Response to Hybrid Threats' conference (2015), 111–12
    Review Magazine, 109, 110
    Strategic Communications Centre of Excellence, Riga, 116, 120
    and subversion-war, 54
    threat to Russia, 164
    Ukrainian Crisis (2014), 107–25
    Very High Readiness Joint Task Force (VJTF), 114, 121, 158
    Wales Summit (2014), 109, 110, 120
North Korea, 28, 63
North Vietnam (1945–76), 63, 70; *see also* Vietnam
novelty of hybrid warfare, 18, 33, 38, 44, 97, 102, 160–63, 164, 166
nuclear weapons, 55, 63–4, 130

Obama, Barack, 109, 124, 174
Odessa, Ukraine, 49, 50, 51
Office of the Assistant Secretary of Defence, 77
Offices of the Institutions of Empress Maria, 50
omnidirectionality, 17, 43
Operation Iraqi Freedom (2003–11), 38, 40–41
Operation Praying Mantis (1988), 36
Orthodox Church, 49, 76

INDEX

Orwell, George, v
Ottoman Empire (1299–1922), 162

pacification, 56
pacifism, 53
Palestine
   al-Aqsa Intifada (2000–2005), 23
   Gaza War (2008–9), 38
Palestine, 70
Panarin, Igor Nikolaevich, 7, 75, 84–90
   and *gibridnaya voyna*, 92, 95, 96, 117, 128, 133, 134, 136, 137, 158
   and meaningful system of management, 88–9
   and political elite, 88
   and private-governmental partnership, 89
   and social objects, 86–7
Parker, Geoffrey, 20
Peace of Westphalia (1648), 19, 57
Peninsular War (1808–14), 26
People's Liberation Army (PLA), 12
Persian Gulf, 36
Philippines, 63
Pinker, Steven, 56, 57
Poland, 108
political warfare, 96–7, 159, 165, 170, 172–3
Potomac Institute for Policy Studies, 37
Potsdam Conference (1945), 62
Prague, Czech Republic, 116
Priebus, Reince, 123
propaganda, 58, 59, 61–2
Protestantism, 76
Putin, Vladimir, 83, 89, 90, 97, 123, 144–6, 148, 174

al-Qaeda, 27, 28, 79–80, 81
Qiao Liang, 12–19, 32, 34, 159, 161
'quagmire' wars, 26

Rácz, Anrás, 112, 114
Radikov, Ivan, 72
RAND Corporation, 38, 119
Rasmussen, Anders Fogh, 109
red scare, 3, 125
Reisinger, Heidi, 107–8
rejection-of-violence (*neprotivlenchestvo*), 55
Renz, Bettina, 103, 114
Riga, Latvia, 116, 120
Robb, John, 31
Roberts, Michael, 20, 25
Rogers, Clifford, 20
Rojansk, Matthew, 167
Romania, 111, 112
Ruiz Palmer, Diego A., 107
Rumsfeld, Donald, 27
Russell, Edward Frederick Langley, 69
Russia and Eurasia Programme, Chatham House, 116
Russian All-Military Union, 52
Russian Civil War (1917–22), 39, 51, 70
Russian Empire (1721–1917), 49–51, 60, 93
   Crimean War (1853–6), 60
   First World War (1914–18), 39, 51, 162
   Japanese War (1904–5), 60
   Napoleonic invasion (1812), 70
   Revolution (1917), 39, 51, 54, 89, 162
   Turkish War (1806–12), 64
   Turkish War (1877–8), 60
Russian Federation, 70–73, 75–90, 91–8, 127–52
   Centre for Military and Strategic Studies, 127
   Chechen Wars (1994–6, 1999–2009), 34, 45, 83, 112, 115, 154
   Crimea annexation (2014), *see under* Crimea annexation

INDEX

Eurasianism, 76, 84, 89
Federal Laws on information, 149
Foreign Policy Concept 147–8
geopolitical interests, 148
Georgian War (2008), 115
Information Security Doctrines, 146–7
information war, 2, 7, 15, 72–3, 75, 84–90, 152
Military Doctrine, 147, 163
Military University, 71, 84, 91
Ministry of Defence, 71, 91, 98
net-centric war, 7, 72, 75–84
oligarchs, 82, 83
presidential elections (2012), 148, 151
sanctions against, 124, 145, 152, 174
Ukrainian Crisis (2014), *see under* Ukrainian Crisis
Russian Liberation Army, 52
Russian Orthodox Church, 49, 76
Russian Revolution (1917), 39, 51, 54, 89, 162, 174
Russian studies, 167
Russo-Georgian War (2008), 115
Russo-Japanese War (1904–5), 60
Russo-Turkish War (1806–12), 64
Russo-Turkish War (1877–8), 60

Saddam Hussein, 80
Sascha-Dominik Bachmann, 105
Savinkin, Alexander, 71
Scheipers, Sibylle, 103
second generation warfare, 19, 20–21, 22
Second Lebanon War (2006), 11, 34–6, 37, 38, 39–40, 45, 102, 154
Second World War (1939–45), 3, 4, 22, 38, 39–40, 52, 62, 165, 174
secularism, 56
Sedan

First battle of (1870), 70
Second battle of (1940), 4
September 11 attacks (2001), 27
Serbia, 51–2
Serebryanikov, Vladimir, 137, 138
sexuality, 53, 54, 56
Shelepin, Leonid, 86
Sherr, James, 172, 175
Sivkov, Konstantin Valentinovich, 94–5
Smith, Adam, 66
Smith, Hanna, 114
Smyslovsky-Holmston, Boris Alexeyevich, 52
social networks, 44
social objects, 86–7
Somalia, 45, 154
Somme, battle of (1916), 4
Sources of Soviet Conduct, The' (Kennan), 87
South African Border War (1966–90), 38
South Korea, 63
Soviet Union (1922–91), 53
  'active measures' (*aktivinye meropriatia*), 3–4, 169
  Afghanistan War (1979–89), 27, 34
  dissolution (1991), *see under* collapse of Soviet Union
  and information war, 87, 89, 117, 171–2
  and Long Telegram (1946), 166, 168–9, 170
  Marxism–Leninism, 71, 148, 170
  and Messner, 49, 70
  network-centric warfare against, 82
  Second World War (1939–45), 39–40
  subversion-war against, 72
  and 'World Revolution' (*Vsemirnaya Revolutsiya*), 56

INDEX

see also Cold War
Spain, 26
St Petersburg, Russia, 50, 51
Stalin, Joseph, 54, 66
Strachan, Hew, 102
Subcommittee on Terrorism, Unconventional Threats and Capabilities, 38
subversion-war, 5, 7, 49–73, 75
   aggressive-diplomacy, 64–6
   Cold War, 49, 52, 54, 62–4, 65, 72, 138
   conservatism, 53, 54, 56, 57, 67
   and *gibridnaya voyna*, 92, 95–7, 117, 127, 128, 132, 133, 134, 136, 137, 155, 158, 159
   half-war, 64–5
   nationalisation of war, 57–9, 60, 65, 67
   psychological dimension, 59–62, 65, 67–8
   and Soviet Union, collapse of (1991), 72
   translation, 66
   'World Revolution' (*Vsemirnaya Revolutsiya*), 55–7, 59, 60, 67
Suez Crisis (1956), 65
Sun Tzu, 162, 171
supra-national powers, 15, 16
Sweden, 3, 25, 105, 144
synchrony, 17
Syria, 36, 39

Taiwan Strait Crisis (1954–5), 65
Taliban, 80
terrorism, 11, 15, 27, 28, 32, 34, 56–7, 70, 79–80, 119, 154
Teufel, Fritz, 69
Thailand, 63
third generation warfare, 19, 21–2
Thirty Years' War (1618–38), 19, 57

Thomas, Timothy, 124
three-wave theory, 55
Toffler, Alvin and Heidi, 55
Tolstoy, Lev, 55
total war, 3, 60
traditional challenges, 28
Treaty of Brest-Litovsk (1918), 62
Trotsky, Leon, 62, 65
Trump, Donald, 122, 123
Tsygankov, Pavel Afanas'yevich, 1, 94
Tyutchev, Fyodor Ivanovich, 166

UAVs (unmanned aerial vehicles), 35
Ukrainian Crisis (2014), 70, 106–25
   and conceptual puzzle, 119, 121
   and democracy, 175
   and force multipliers, 117
   and *gibridnaya voyna*, 92
   as hybrid warfare, 2, 94, 96, 97, 101, 106–25, 139, 151, 154–7, 160
   as international relations problem, 124, 125
   and information warfare, 144–5
   NATO reaction to, 101, 107–25, 154
   politicisation, 118–23, 125
   public views on, 152
   and Russian studies, decline in, 168
United Kingdom, 76, 87, 90, 111, 122, 162
United Nations, 16
United States, 7, 11–29, 31–45
   'active measures' against, 3–4, 169
   Afghanistan War (2001–14), 112, 119, 160
   Civil War (1861–5), 21
   Cold War, *see under* Cold War
   Command and Control Research Program (CCRP), 77
   compound warfare, 12, 24–7, 31, 32–3, 36, 154, 158, 159

INDEX

Democratic National Committee email leak (2016), 118
Department of Defense (DoD), 27–9, 45
  Dugin's views on, 76
  fourth generation warfare, *see under* fourth generation warfare
  and *gibridnaya voyna*, 96, 97, 133, 137, 138, 139, 149, 157, 165, 169, 170
Government Accountability Office, 38
Gulf War (1990–91), 12, 16–17, 129
House of Representatives, 38
hybrid warfare, *see under* Hoffman, Frank
information war, use of, 133, 137, 139, 145–52, 153
Iraq War (2003–11), 38, 40–41, 112, 160
Joint Forces Command Joint Irregular Warfare Centre (USJFCOM JIWC), 104–5
Kosovo War (1998–9), 129
Long Telegram (1946), 166, 168–9, 170
Marine Corps (USMC), 19, 22, 32, 37, 102
military culture, 42–3, 103
National Defence Strategy (2005), 12, 27–9, 33, 154
National Defense University (NDU), 104
network-centric warfare, 76–7, 79, 81–4
Office of the Assistant Secretary of Defence, 77
Operation Praying Mantis (1988), 36
political warfare, 96–7, 159, 165, 170, 172–3
presidential election (2016), 118, 122, 123, 172

red scare, 3, 125
Revolutionary War (1775–83), 27, 65, 162
September 11 attacks (2001), 27
  and subversion-war, 54
  Taiwan Strait Crisis (1954–5), 65
  and technology, 43
  unrestricted warfare, 12–19, 31, 36, 37, 66, 154, 158
  Vietnam War (1965–73), 23, 27, 40, 69–70, 112
unrestricted warfare, 12–19, 31, 32, 36, 37, 66, 154, 158, 159
urbanisation, 56

van Creveld, Martin, 19
van Kappen, Frank, 108
Verdun, battle of (1916), 23
Very High Readiness Joint Task Force (VJTF), 114, 121, 158
Vietnam
  First Indochina War (1946–54), 65, 66
  Second Indochina War (1955–75), 23, 27, 40, 63, 69–70, 112
Vladimirov, Alexander, 72
Vlasov, Andrey, 52
Volker, Kurt, 108
Volkovsky, Nikolay, 130

Wales NATO Summit (2014), 109, 110, 120
Wang Xiangsui, 12–19, 32, 34, 159, 161
war, misuse of term, 156, 157–8
Warsaw Pact, 119, 137
Waterloo, battle of (1815), 23
White Movement, 51, 52, 70, 71
Williamson, Steven, 38
Wilson Center, 167
World Bank, 16, 81

INDEX

World Council of Churches, 69
'World Revolution' (*Vsemirnaya Revolutsiya*), 55–7, 59, 60
World Trade Organisation (WTO), 16, 81
Wrangel, Pyotr Nikolayevich, 51, 52
Württemberg, Germany, 49

Yugoslav Wars (1991–2001), 34, 70, 129, 134, 137

Zaborowski, Marcin, 108
Zhou Dynasty (c. 1046 BC–256 BC), 19
Zyuganov, Gennady, 151